Evaluating the Effectiveness of Offshore Safety and Environmental Management Systems

Committee on the Effectiveness of Safety
and Environmental Management Systems
for Outer Continental Shelf Oil
and Gas Operations

TRANSPORTATION RESEARCH BOARD
OF THE NATIONAL ACADEMIES

Transportation Research Board
Washington, D.C.
2012
www.TRB.org

Transportation Research Board Special Report 309

Subscriber Categories
Safety and human factors, marine transportation, environment

Transportation Research Board publications are available by ordering individual publications directly from the TRB Business Office, through the Internet at www.TRB.org or national-academies.org/trb, or by annual subscription through organizational or individual affiliation with TRB. Affiliates and library subscribers are eligible for substantial discounts. For further information, contact the Transportation Research Board Business Office, 500 Fifth Street, NW, Washington, DC 20001 (telephone 202-334-3213; fax 202-334-2519; or e-mail TRBsales@nas.edu).

Copyright 2012 by the National Academy of Sciences. All rights reserved.
Printed in the United States of America.

NOTICE: The project that is the subject of this report was approved by the Governing Board of the National Research Council, whose members are drawn from the councils of the National Academy of Sciences, the National Academy of Engineering, and the Institute of Medicine. The members of the committee responsible for the report were chosen for their special competencies and with regard for appropriate balance.

This report has been reviewed by a group other than the authors according to the procedures approved by a Report Review Committee consisting of members of the National Academy of Sciences, the National Academy of Engineering, and the Institute of Medicine.

This study was sponsored by the Bureau of Safety and Environmental Enforcement of the U.S. Department of the Interior.

Library of Congress Cataloging-in-Publication Data

United States. Committee on the Effectiveness of Safety and Environmental Management Systems for Outer Continental Shelf Oil and Gas Operations.
  Evaluating the effectiveness of offshore safety and environmental management systems/ Committee on the Effectiveness of Safety and Environmental Management Systems for Outer Continental Shelf Oil and Gas Operations.
      pages cm.—(Transportation Research Board special report ; 309)
   Includes bibliographical references.
   ISBN 978-0-309-22308-9
   1. Offshore oil well drilling—United States—Safety measures—Evaluation. 2. Offshore gas well drilling—United States—Safety measures—Evaluation. 3. Continental shelf—United States. 4. Environmental protection—Government policy—United States.
I. National Research Council (U.S.). Transportation Research Board. II. Title.
   TN872.A45 2012
   363.11'9622338190973—dc23
                              2012026029

THE NATIONAL ACADEMIES
Advisers to the Nation on Science, Engineering, and Medicine

The **National Academy of Sciences** is a private, nonprofit, self-perpetuating society of distinguished scholars engaged in scientific and engineering research, dedicated to the furtherance of science and technology and to their use for the general welfare. On the authority of the charter granted to it by the Congress in 1863, the Academy has a mandate that requires it to advise the federal government on scientific and technical matters. Dr. Ralph J. Cicerone is president of the National Academy of Sciences.

The **National Academy of Engineering** was established in 1964, under the charter of the National Academy of Sciences, as a parallel organization of outstanding engineers. It is autonomous in its administration and in the selection of its members, sharing with the National Academy of Sciences the responsibility for advising the federal government. The National Academy of Engineering also sponsors engineering programs aimed at meeting national needs, encourages education and research, and recognizes the superior achievements of engineers. Dr. Charles M. Vest is president of the National Academy of Engineering.

The **Institute of Medicine** was established in 1970 by the National Academy of Sciences to secure the services of eminent members of appropriate professions in the examination of policy matters pertaining to the health of the public. The Institute acts under the responsibility given to the National Academy of Sciences by its congressional charter to be an adviser to the federal government and, on its own initiative, to identify issues of medical care, research, and education. Dr. Harvey V. Fineberg is president of the Institute of Medicine.

The **National Research Council** was organized by the National Academy of Sciences in 1916 to associate the broad community of science and technology with the Academy's purposes of furthering knowledge and advising the federal government. Functioning in accordance with general policies determined by the Academy, the Council has become the principal operating agency of both the National Academy of Sciences and the National Academy of Engineering in providing services to the government, the public, and the scientific and engineering communities. The Council is administered jointly by both Academies and the Institute of Medicine. Dr. Ralph J. Cicerone and Dr. Charles M. Vest are chair and vice chair, respectively, of the National Research Council.

The **Transportation Research Board** is one of six major divisions of the National Research Council. The mission of the Transportation Research Board is to provide leadership in transportation innovation and progress through research and information exchange, conducted within a setting that is objective, interdisciplinary, and multimodal. The Board's varied activities annually engage about 7,000 engineers, scientists, and other transportation researchers and practitioners from the public and private sectors and academia, all of whom contribute their expertise in the public interest. The program is supported by state transportation departments, federal agencies including the component administrations of the U.S. Department of Transportation, and other organizations and individuals interested in the development of transportation. **www.TRB.org**

www.national-academies.org

Committee on the Effectiveness of Safety and Environmental Management Systems for Outer Continental Shelf Oil and Gas Operations

Kenneth E. Arnold, WorleyParsons, Inc., Houston, Texas, *Chair*
J. Ford Brett, PetroSkills, Tulsa, Oklahoma
Paul S. Fischbeck, Carnegie Mellon University, Pittsburgh, Pennsylvania
Stuart Jones, Lloyd's Register EMEA, Aberdeen, Scotland, United Kingdom
Thomas Kitsos, Consultant, Bethesda, Maryland
Frank J. Puskar, Energo Engineering, Houston, Texas
Darin W. Qualkenbush, Chevron North America Exploration and Production Company, Covington, Louisiana
Raja V. Ramani, Pennsylvania State University, University Park, Pennsylvania (emeritus)
Vikki Sanders, JMJ Associates, Austin, Texas

Transportation Research Board Staff
Beverly Huey, Project Director

Preface

Although relatively few significant incidents have occurred on oil and gas installations worldwide in recent years, those that have occurred (especially the Macondo Well incident in April 2010) have underscored the need to enhance the effectiveness of inspection programs for offshore installations. From its inception in 1982 until October 2010, the Minerals Management Service (MMS) of the U.S. Department of the Interior was the responsible regulatory authority for the offshore oil and gas industry in U.S. waters; during this period its role continued to develop as technologies, expectations, and guidelines for safe and environmentally friendly operations evolved.

In the late 1980s, MMS approached the Marine Board of the National Research Council (NRC) "to develop inspection strategies to improve safety and the effectiveness of the inspection process" (NRC 1990, vi). The committee that was formed was tasked with reviewing the current inspection program for the Outer Continental Shelf (OCS), appraising inspection practices elsewhere, developing alternatives for conducting inspection programs and assessing their advantages and disadvantages, and recommending alternative inspection procedures that might be more effective and efficient.

Following the release of that report, the industry was encouraged to adopt safety and environmental management programs voluntarily. At the same time, MMS began examining its regulatory oversight and, in mid-2009, proposed a rule that would have required offshore operators[1]

---

[1] An operator is defined as "The individual, partnership, firm, or corporation having control or management of operations on the leased area or a portion thereof. The operator may be a lessee, designated agent of the lessee(s), or holder of operating rights under an approved operating agreement" (API 2004, Appendix D).

to adopt four of the 12 elements of American Petroleum Institute (API) Recommended Practice (RP) 75 (API 2004).

In April 2009, MMS again approached the Marine Board to request that a study be conducted to review the MMS inspection program for offshore facilities to assess its effectiveness in protecting human safety and the environment. The Committee on Offshore Oil and Gas Facilities Inspection Program of the MMS (which was later renamed the Committee on the Effectiveness of Safety and Environmental Management Systems for Outer Continental Shelf Oil and Gas Operations) was tasked with

- Examining changes in the inspection program and process since the 1990 Marine Board study;
- Reviewing available trend data on inspections, safety, and environmental damage;
- Examining analogous safety inspection programs in other regulatory agencies and other nations for lessons that could be applied to MMS inspections;
- Considering both the changes in the industry's safety management practices since the 1990 Marine Board report and the implications of these changes for MMS inspection practices;
- Considering the effects of the current inspection program on offshore safety and environmental protection; and
- Recommending changes, as appropriate, to the inspection program to enhance effectiveness.

The committee includes members, practitioners, and academicians who bring a broad spectrum of expertise that includes the areas of safety management, human factors, risk assessment, organizational management and management systems, offshore engineering, offshore platform design and construction, offshore operations, and policy as well as the areas of safety regulations and inspections in related industries. It was appointed in November 2009, held its first meeting the following month, and conducted site visits in March 2010 to the Pacific OCS region and to the California State Lands Commission. The committee also scheduled offshore site visits in May of that same year to the MMS Gulf of Mexico region. Those visits, however, were overtaken by the unfolding events of the April 2010 *Deepwater Horizon*–Macondo well blowout, at which time MMS officials requested that this project be put on hold while the agency reevaluated its approach to safety.

Immediately after the Macondo well blowout, many investigations and inquiries were launched and far-reaching changes in the U.S. offshore oil and gas industry were initiated. The Department of the Interior undertook a major reorganization of MMS that initially separated the agency's revenue management functions from its nonfiscal responsibilities (e.g., oversight and regulatory enforcement functions) and later separated the ocean energy management functions from the safety and environmental enforcement functions. In addition, the industry itself began discussing plans to develop an independent organization to work with both industry and the regulators to enhance the effectiveness of safety and environmental regulations and programs.

In October 2010, the Bureau of Ocean Energy Management, Regulation and Enforcement (BOEMRE) (formerly MMS) issued a final rule, Safety and Environmental Management Systems (SEMS) (BOEMRE 2010), which required adoption of API RP 75. In the SEMS rule, BOEMRE recognized that its inspection program was too focused on mechanical failures and that such failures represent a small minority of incidents. With the issuance of the final rule, BOEMRE expanded its approach to safety and environmental protection to encompass not only reliance on inspection of hardware-oriented items related to potential incidents of noncompliance, but also safety management. Operators were required to specify how they would manage safety holistically to avoid injury and spills.

In October 2011, BOEMRE was replaced by the Bureau of Ocean Energy Management (BOEM) and the Bureau of Safety and Environmental Enforcement (BSEE). BSEE has broad regulatory authority over energy operations on the OCS, including oversight responsibility with respect to the offshore installations involved in drilling and production of oil and natural gas. Included in BSEE's oversight authority is the responsibility for conducting inspections and audits. It is expected that the audit process will encourage owners and operators to develop a safe and environmentally friendly operational process on offshore facilities and that, if there are potential problems, these will be identified during the audit process and subsequently addressed, thereby reducing the likelihood of a major incident.

Following the restructuring of MMS and issuance of the final rule in late 2010, BOEMRE requested that the scope of the present study be changed from a review of the agency's prior offshore facility safety and environmental inspection program to one that provided guidance on how the agency should evaluate and ensure the effectiveness of the implementation of the new SEMS practices that were to be required of offshore operators as of November 15, 2011. As a consequence, this project was refocused, and the committee resumed its work in late January 2011.

Under the new agreement with BOEMRE, the Committee on the Effectiveness of Safety and Environmental Management Systems for Outer Continental Shelf Oil and Gas Operations was tasked with preparing an interim report that would identify potential methods for assessing the effectiveness of a company's SEMS program and describe the pros and cons of each method as they were known to that point. After the committee resumed its activities, it met four times in 2011: March 3 and 4, August 31 and September 1, October 19 and 20, and December 1. A subgroup of the committee also attended the BOEMRE-sponsored Public Workshop on Offshore Energy Regulations on March 15, 2011, to keep abreast of interpretation and proposed implementation of the SEMS regulation (30 CFR 250, Subpart S).

The committee released its interim report on June 28, 2011 (TRB 2011). After the October committee meeting, another subgroup of the committee visited offshore facilities in the Gulf of Mexico region to complete the data-gathering process.

This final report describes the various methods that BSEE may employ to evaluate the effectiveness of an operator's SEMS program, recommends a holistic method that the committee believes should be adopted, and provides guidance on how this goal can be accomplished. The committee would like to note that some of the efforts that BSEE and the industry have already undertaken in the aftermath of the Macondo well incident are supported and reinforced by the recommendations in this report. These are steps in the right direction that need to be built on in a timely manner to ensure that operational and environmental risks are reduced.

ACKNOWLEDGMENTS

The work of the committee was greatly facilitated by the thoughtful advice and background information provided by all of the presenters at its meetings, other individuals with relevant technical expertise, and government and industry officials who were consulted during the course of the study. The committee also gratefully acknowledges the contributions of time and information provided by the sponsor liaisons: Elmer Danenberger (ret.), Douglas Slitor (ret.), David Nedorostek, and Staci King.

To obtain a better understanding of alternative regulatory approaches, the committee also received presentations and briefings from Mike Marshall, Process Safety Management Coordinator, U.S. Department of Labor–Occupational Safety and Health Administration Directorate of Enforcement Programs; Linda Zeiler, Acting Director, Technical Support, Mine Safety and Health Administration; Magne Ognedal, Director General, Petroleum Safety Authority Norway; Chet Miller, Inspector, California District, Pacific OCS Region, BSEE; James Munro, Operations Manager, Hazardous Installations Directorate, Offshore Division, United Kingdom Health and Safety Executive; Jeff Scott, SempCheck Services; Ian Sutton, Process Risk Management, AMEC Paragon; Anton du Preez, SempCheck Services; Jack Toellner, Senior Technical Advisor—Safety, ExxonMobil Development Company; Allen J. Verret, Executive Director, Offshore Operators Committee; Charlie Williams, Well Delivery Manager and Chief Scientist, Upstream Americas, Shell; Kyle Wingate, Principal Human Factors Consultant, Scandpower, Inc., Lloyd's Register Group; Peter Masters, Product Manager, Lloyd's Register Group Services Limited; and Luiz Feijo, Manager, Project Development, ABS.

In addition, the committee thanks the many industry, trade association, and state and local government representatives and other individuals who provided input for this study. In particular, the committee would like to thank the staff at the MMS (now BSEE) Pacific OCS regional office—Phillip Schroeder, District Manager, California District; Rishi Tyagi, Regional Supervisor, Office of Field Operations; Ralph Vasquez, Supervisory Inspector; Chet A. Miller, Inspector; and Ellen Aronson, Regional Directorate—and at the U.S. Coast Guard, Santa Barbara—Lt. Chad J. Robuck, Supervisor, Marine Safety Detachment; and CWO4 J. P. Giles, Senior Marine Inspector. The committee also appreciates having had the opportunity to meet with staff at the Mineral Resources Management Division, California State Lands Commission: Mark Steinhilber, Senior Process Safety Engineer; David Rodriguez, Associate Process Safety Engineer; Peter V. Johnson, Operations Manager; and Jack L. Smith, Senior Mineral Resources Engineer.

The committee also thanks the crew of Platform Gail off the coast of California—especially Tony Martinez, Operations Foreman; and Chris, Facilities Engineer—and the crew of Genesis Platform in the Gulf of Mexico—especially Eric Farmer, Safety Management Systems Medic; Mervin Broussard, Chevron Field Coordinator; and Joe St. Ann, Offshore Installation Manager—who welcomed committee members onboard their installations, as well as those who arranged these offshore visits, particularly David Dykes, Douglas Slitor, and Ralph Vasquez.

This study was performed under the overall supervision of Stephen R. Godwin, Director of Studies and Special Programs, Transportation Research Board (TRB). The committee gratefully acknowledges the work and support of Beverly Huey, who served as project director and under whose guidance this study was initiated and undertaken. The committee also acknowledges the work and support of Suzanne Schneider, Associate Executive Director of TRB, who managed the review process; Janet M. McNaughton, who edited the report and handled the editorial production; Juanita Green, who managed the production; Jennifer J. Weeks, who prepared the manuscript for prepublication web posting; and Javy Awan, Director of Publications, under whose supervision the report was prepared for publication. Claudia Sauls, Amelia Mathis, and Mai Q. Le arranged meetings and provided administrative support to the committee.

This report was reviewed in draft form by individuals chosen for their diverse perspectives and technical expertise, in accordance with procedures approved by the NRC Report Review Committee. The purpose of this independent review is to provide candid and critical comments that will assist the institution in making its published report as sound as possible and to ensure that the report meets institutional standards for objectivity, evidence, and responsiveness to the study charge. The review comments and draft manuscript remain confidential to protect the integrity of the deliberative process.

The committee thanks the following individuals for their review of this report: Richard Hartley, B&W Pantex, Amarillo, Texas; Morgan L. Jones, PetroSkills, Tulsa, Oklahoma; Najmedin Meshkati, University of Southern California, Los Angeles; Ali Mosleh, University of Maryland, College Park; Walt Rosenbusch, International Association of Geophysical Contractors, Houston, Texas; Mark Steinhilber, California State Lands Commission, Long Beach, California; Stanley C. Suboleski, Virginia Polytechnic Institute and State University (ret.), Blacksburg; and Allen J. Verret, Offshore Operators Committee, Metairie, Louisiana. Although these reviewers provided many constructive comments and suggestions, they were not asked to endorse the committee's conclusions or recommendations, nor did they see the final draft of the report before its release.

The review of this report was overseen by Hyla S. Napadensky, Grand Marais, Minnesota, and C. Michael Walton, University of Texas at Austin. Appointed by NRC, they were responsible for making certain that an independent examination of this report was carried out in accordance with institutional procedures and that all review comments were carefully considered. Responsibility for the final content of this report rests entirely with the authoring committee and the institution.

> Kenneth E. Arnold, *Chair*
> Committee on the Effectiveness of Safety
> and Environmental Management Systems for
> Outer Continental Shelf Oil and Gas Operations

REFERENCES

Abbreviations

API	American Petroleum Institute
BOEMRE	Bureau of Ocean Energy Management, Regulation, and Enforcement
NRC	National Research Council
TRB	Transportation Research Board

API. 2004. *Recommended Practice for Development of a Safety and Environmental Management Program for Offshore Operations and Facilities,* 3rd ed. API RP 75. API, Washington, D.C.

BOEMRE. 2010. Oil and Gas and Sulphur Operations in the Outer Continental Shelf—Safety and Environmental Management Systems. *Federal Register,* Vol. 75, No. 199, Fri., Oct. 15, 2010, pp. 63610–63654.

NRC. 1990. *Alternatives for Inspecting Outer Continental Shelf Operations.* National Academies Press, Washington, D.C.

TRB. 2011. Interim Report on the Effectiveness of Safety and Environmental Management Systems for Outer Continental Shelf Oil and Gas Operations. http://onlinepubs.trb.org/onlinepubs/sr/srSEMSInterimReport.pdf.

Acronyms

Acronyms used in the report are listed below.

ABS	American Bureau of Shipping
ALARP	as low as reasonably practicable
API	American Petroleum Institute
BOEM	Bureau of Ocean Energy Management
BOEMRE	Bureau of Ocean Energy Management, Regulation and Enforcement
BSEE	Bureau of Safety and Environmental Enforcement
CFR	*Code of Federal Regulations*
CIP	continuous improvement program
COS	Center for Offshore Safety
CSLC	California State Lands Commission
DOI	Department of the Interior
DDRS	Daily Drilling Report System
HSE	Health and Safety Executive (United Kingdom)
INC	incident of noncompliance
ISM	International Safety Management
ISO	International Organization for Standardization
JSA	job safety analysis
KPI	key performance indicator
MMS	Minerals Management Service
MOC	management of change
MSHA	Mine Safety and Health Administration
NIOSH	National Institute for Occupational Safety and Health
NSSGA	National Stone, Sand, and Gravel Association
NVIC	Navigation and Vessel Inspection Circular

OCS	Outer Continental Shelf
OSHA	Occupational Safety and Health Administration
PHA	process hazard analysis
PINC	potential incident of noncompliance
PSA	Petroleum Safety Authority (Norway)
PSM	process safety management
QRA	quantitative risk assessment
RNNP	*Risikonivå i norsk petroleumsvirksomhet*
RP	Recommended Practice
SAMS	Safety Assessment of Management Systems
SEMP	safety and environmental management program
SEMS	Safety and Environmental Management Systems
SME	Society for Mining, Metallurgy, and Exploration
SMS	safety management system
SPE	Society of Petroleum Engineers
UK	United Kingdom
USCG	U.S. Coast Guard
USCOP	U.S. Commission on Ocean Policy

Contents

Summary	1
1 Introduction	**11**
Study Context	12
Study Objective and Charge	16
Organization of the Report	17
2 Role of Safety and Environmental Management Systems in Establishing a Culture of Safety	**18**
Will SEMS Promote a Culture of Safety?	19
Guiding Questions for Evaluation or Audit	26
Assessing the Effectiveness of SEMS and Its Effect on Culture	27
A Word of Hope	29
3 Methods for Assessing Effectiveness	**32**
Potential Assessment Methods	32
Measuring Trends	41
Summary	41
4 Existing Approaches for Assessing Safety Management Systems	**43**
U.S. Regulatory Agencies	44
International Regulatory Organizations	55
Center for Offshore Safety: A Self-Policing Safety Organization	67
Summary	70

5 Role of the Bureau of Safety and Environmental Enforcement in Evaluating Safety and Environmental Management Systems Programs	**72**
Inspections	72
Audits	74
Ensuring Effectiveness	85
6 Conclusions and Recommended Approach	**89**
Conclusions	89
Recommended Approach	94
Resources Required	105
References	**106**
Study Committee Biographical Information	**111**

Summary

For many years the United States employed a prescriptive regulatory system for the offshore oil and gas industry in which operators were required to demonstrate conformance with established regulations. In the aftermath of the April 2010 Macondo well blowout and explosion, the federal government and the offshore oil and gas industry have been undergoing major changes, including the issuance of regulations requiring operators of offshore oil and gas facilities to adopt and implement comprehensive Safety and Environmental Management Systems (SEMS) programs.

SEMS is a safety management system (SMS) aimed at shifting from a completely prescriptive regulatory approach to one that is proactive, risk based, and goal oriented in an attempt to improve safety and reduce the likelihood that events similar to the Macondo incident will reoccur. Although the new regulations had been voluntary for many years and a subset of these components had been proposed in rulemaking before the Macondo well accident, it was not until this major accident that comprehensive changes were made. The Committee on the Effectiveness of Safety and Environmental Management Systems for Outer Continental Shelf Oil and Gas Operations (the committee), which conducted the present study, was charged with recommending a method of assessing the effectiveness of operators' SEMS programs on any given offshore drilling or production facility.

Safety professionals have understood for decades that to increase safety in complex industrial installations, organizations must manage safety with the same principles of planning, organization, implementation, and investigation that they use to carry out any other business function. In 1992 the federal government promulgated a process safety management

(PSM) regulatory approach for installations that handle highly hazardous chemicals. PSM specified the elements that must be included in a plan to manage safety. A similar risk management approach was mandated for facilities handling certain chemicals regulated under the Clean Air Act Amendments of 1996. In parallel, the American Petroleum Institute (API) developed Recommended Practice (RP) 75, *Recommended Practice for Development of a Safety and Environmental Management Program for Offshore Operations and Facilities* (API 1993, 2004).

The federal government initially encouraged the offshore oil and gas industry to adopt API RP 75 voluntarily and from 1994 to 1998 used a self-report survey to monitor the level of adoption of each element. After reviewing the analysis and comments received in response to a 2006 advance notice of proposed rulemaking to make portions of API RP 75 mandatory, the Minerals Management Service (MMS) proposed to require each offshore lessee–operator to develop, implement, maintain, and operate a SEMS program that contained four elements of API RP 75: hazards analysis, management of change, operating procedures, and mechanical integrity. The Offshore Operators Committee and others recommended, however, that if a SEMS rule were to become mandatory, it should include all of the elements of a safety and environmental management program (SEMP) discussed in API RP 75, and not just the four listed for the proposed rule. MMS was preparing a rule to require the implementation of all of the SEMP elements in API RP 75 when the occurrence of the Macondo well blowout delayed publication of the new rule. The final SEMS rule was promulgated in the *Federal Register* on October 15, 2010 (BOEMRE 2010) and became effective on November 15, 2011 (30 CFR 250, Subpart S).

Mandating SEMS programs and ensuring their effectiveness is a step toward improving governmental oversight of the offshore oil and gas industry and industry implementation of reforms to reduce the risk of accidents and to improve safety, which is needed according to some of the investigations of the Macondo well blowout (e.g., NAE-NRC 2011; National Commission on the BP *Deepwater Horizon* Oil Spill and Offshore Drilling 2011). The committee agrees with these conclusions.

The 1990 Marine Board study *Alternatives for Inspecting Outer Continental Shelf Operations* made the crucial point that the emphasis that regulators and the industry had placed on compliance with very specific regulations and rigid checklists to ensure compliance was not the best way to change attitudes toward safety (NRC 1990). In enterprises that are subject to checklist-style compliance inspections by government authorities, passing the inspection comes to be seen as equaling safety. This compliance mentality does not necessarily correlate with an increase in the level of safety attitudes and actions on the part of the companies and individuals involved in the actual operations.

Instillation of an appropriate culture of safety in an operation requires mechanisms that

- Establish structure and control by specifying what is needed for safe operation and checking to see that these specifications are being followed, and
- Build competency by developing individual knowledge and skill.

In addition to these mechanisms, there must be actions that establish norms and motivations that encourage those who are making decisions to constantly want to think about safety and behave in ways that maximize safety. Thus, whereas having an adequately functioning SEMS-type program is necessary to develop an appropriate culture of safety, SEMS by itself is not sufficient. To be successful, the tenets of SEMS must be fully acknowledged and accepted by workers, motivated from the top, and supported throughout the organization and must drive workers' actions; only then can an effective culture of safety be established and grow.

The committee believes that the approach ultimately taken by the Bureau of Safety and Environmental Enforcement (BSEE) to evaluate the effectiveness of SEMS can have a positive impact on norms and motivations and move both BSEE and the industry away from a compliance mentality to one that encourages an ever-evolving and -improving culture of safety in offshore operations. To encourage a culture of safety in which individuals know the safety aspects of their actions and are motivated to think about safety, the agency will need to adopt and evolve an evaluation system for SEMS that emphasizes the assessment of attitudes and actions rather than documentation and paperwork. All of the elements of SEMS must be addressed, but it is more important that those who are actually doing the work understand and practice these elements than that these elements be documented. BSEE should look beyond its predecessor agencies' historical role of assuring compliance with prescriptive regulations and seize the current opportunity to redefine its role, at least partially, to one of encouraging an atmosphere that helps the industry migrate from a compliance mentality to a culture of safety that includes compliance. Furthermore, an organization's SEMS program must incorporate a dynamic process that evolves with time; thus, to be effective, the procedures, inspections, and audits employed by BSEE to verify the effectiveness of an operator's SEMS program should also be dynamic. Likewise, inspection and audit criteria will need to be dynamic so that they do not become outdated as new technologies are employed and new environments explored.

The purpose of this report is to define the broad outlines of a holistic approach BSEE can take to evaluate the effectiveness of operators' SEMS programs. It is not possible, however, for a regulator to create a culture of safety in an organization by inspection or audit; that culture needs to come from within the organization. The regulator's role is to regulate in a manner that helps the organization be safe.

SEMS, by definition, is a program for managing the overall safety and environmental aspects of an offshore oil and gas operation. Unfortunately, no single, existing set of statistics can measure the effectiveness of SEMS on an offshore installation. Certainly there are statistics such as fatality rates, injury rates, and lost-time incidents that correlate with the level of what is often referred to as "personal safety" or "worker safety" incidents. It is much harder, if not impossible, to identify similar statistics that correlate with what the Occupational Safety and Health Administration calls "process safety" and what the National Academy of Engineering and National Research Council Committee for the Analysis of Causes of the *Deepwater Horizon* Explosion, Fire, and Oil Spill to Identify Measures to Prevent Similar Accidents in the Future (NAE-NRC 2011) calls "system safety" (i.e., the possibility of the occurrence of a very low-probability, very high-consequence event such as the Macondo well blowout). Ensuring the effectiveness of SEMS for both worker safety and system safety will depend on a thorough commitment by industry and government application of best practices applied by other effective regulators.

In recommending a holistic approach to evaluating the effectiveness of SEMS programs, the committee discussed in detail SEMS' role in helping to develop a culture of safety, looked at the pros and cons of various methods of assessing the effectiveness of a SEMS program, and investigated existing approaches for assessing the SMS programs of various U.S. and international regulatory agencies whose safety mandates are similar to that of BSEE. The committee received presentations from, and conducted follow-up inquiries with, the Occupational Safety and Health Administration and the Mine Safety and Health Administration of the U.S. Department of Labor, the U.S. Coast Guard, and the California State Lands Commission (CSLC) as well as with the United Kingdom Health and Safety Executive and Petroleum Safety Authority (PSA) Norway.

It seems clear from the experiences of other regulatory agencies, especially CSLC and PSA Norway, that other organizations with many years of experience in overseeing SEMS-like programs have migrated toward a system that

- Audits operations with a qualified team of auditors,
- Assesses through discussions with personnel at different levels of the operation the way in which the elements of the SMS are actually being used,
- Feeds the results back to the top management of the operating companies, and
- Expects continuous improvement and monitors for it.

These agencies have found that engagement with the industry is more productive than punishment, although they maintain the threat of punishment if needed.

RECOMMENDED APPROACH

On the basis of the information obtained from presentations to the committee, site visits, published regulations, notices of proposed rulemaking, API-recommended practices, and previously published reports, **the committee recommends that BSEE take a holistic approach to evaluating the effectiveness of SEMS programs. This approach should, at a minimum, include inspections, audits (operator and BSEE), key performance indicators, and a whistleblower program.**

Inspections

BSEE should continue its current program of ensuring compliance with specific regulations. The routine presence of competent BSEE inspectors on an offshore operator's facility should be used to verify that the industry is generally complying with SEMS. Without proper training, however, BSEE inspectors will have a tendency to issue incidents of noncompliance (INCs) for deviations of documentation from a checklist, and such deviations may or may not be important in meeting the intent of SEMS. In turn, the issuing of INCs may focus operator attention on compliance in the way documentation is written rather than on establishing a culture that actually promotes safety. Therefore, BSEE should train inspectors to employ other options in addition to issuing citations. BSEE inspectors should look beyond the written regulation to identify operators in marginal compliance and guide them into a more complete state of compliance. In doing so, BSEE inspectors could help focus BSEE-initiated SEMS audits (see below).

Making judgments about organizational safety culture and SEMS compliance will require training inspectors and scheduling of inspections to allow inspectors to spend more time offshore interacting with operating staff and observing day-to-day operations. The necessary resources could come from other sources, including the use of operator-provided transportation and accommodations or from an increase in inspection fees. Other regulatory organizations use operator-furnished transportation and accommodations with no adverse effect on the integrity of the process. BSEE should consider doing the same to increase the quality of its inspections and to reduce expenditures. In addition to providing a financial benefit, the use of operator-furnished transportation and accommodations will help achieve the goal of greater informal interaction between inspectors and operating staff and will aid inspectors in making a better evaluation of the level of safety that exists.

Audits

Operator Audits

It is critical that SEMS programs be audited. The frequency of the audits should be risk based. Annual audits may be necessary for very large installations, while other, noncritical installations may not require specific audits beyond normal inspection observations. Audits should be carried out by the operator's internal qualified, independent team wherever possible. Operator responsibility for audits will help prevent the development of a compliance mentality. Smaller operations that may find it necessary to use third-party auditors should include on the audit team at least one operator employee who is not directly involved in the day-to-day operations of the installation being audited. In cases in which meeting this requirement is not possible (e.g., very small operators with only a handful of employees) it may be necessary for the chief executive officer of the company to participate as a member of the audit team. Nevertheless, BSEE should approve all audit plans to ensure adequate frequency of auditing and the quality of the proposed audit team. BSEE should also receive a copy of each audit and follow-up report.

A truly independent internal audit team is preferred to an external, third-party team. Use of a well-documented internal team would help to ensure a quality audit that also encourages an appropriate culture of safety. BSEE, in consultation with the industry and, potentially, the Center for Offshore Safety, should develop an approach to certify auditors, develop audit standards, and establish the process by which audits themselves are conducted.

BSEE Audits

BSEE should perform complete or partial audits of SEMS programs when justified by reports from inspectors, reviews of operators' audit reports, incidents, or events. BSEE is responsible for verifying that quality audits are carried out and acted on appropriately. Because of the comprehensive nature of the SEMS requirements, BSEE's oversight of internal and third-party audits needs to include a range of techniques, each of which focuses on a different aspect of an operation's safety system. BSEE can use reports from its compliance inspectors and its reviews of audit reports to identify the need for specific BSEE-conducted targeted or spot audits, or complete audits, to determine whether an operator's SEMS program is improving safety. Interviews, demonstrations, and observations, rather than checklists, are necessary to make such a determination. To perform these audits and review operator audit plans and internal audits, BSEE needs a cadre of trained auditors who will be able to spend sufficient time on location to conduct the appropriate audits. Hiring and training additional personnel will most likely be necessary.

Key Performance Indicators

Over time and in consultation with other national and international regulatory bodies that collect similar data, BSEE should also develop key performance indicators or other indicators that could be useful in providing a measure of the effectiveness of an operator's or offshore installation's SEMS program and culture of safety. BSEE can collect and evaluate data from operations within and across platforms to identify specific problems and trends in operations at a particular facility and across the industry. This information is also needed to evaluate the SEMS audit approach and to identify opportunities for improvement. Because BSEE will review all audit and follow-up reports in addition to having access to inspections and its own audits, the agency will be in the best position to disseminate findings and best practices of a general interest.

Whistleblower Program

BSEE should establish a whistleblower program to help monitor the culture of safety that actually exists at each installation and to help uncover any improprieties in its own operations. Workers must have a way to anonymously report not only dangerous deviations in norms and motivations that may not be obvious to BSEE inspectors or even to internal auditors, but also unprofessional conduct by BSEE's own staff. Care should be taken in devising this program to make sure that it does not become a tool for disgruntled employees seeking to punish perceived wrongs.

CONCLUDING COMMENTS

In the immediate aftermath of the *Deepwater Horizon* blowout, the U.S. Department of the Interior initiated a major restructuring (and separation of conflicting responsibilities) of the former MMS, as well as sweeping reforms in regulatory oversight of the offshore oil and gas industry. These changes have begun to change the industry's approach to safety management and will, it is believed, reduce risk and result in positive changes in the industry's culture of safety.

The committee examined the new regulations that were promulgated in October 2010 and went into effect November 15, 2011. By early September 2011, during the writing of this report, BSEE published in the *Federal Register* a notice of proposed changes in SEMS (BOEMRE 2011). All but one of these proposed changes are consistent with the findings of this report. The one change that is not requires that SEMS audits be performed by independent third parties. This committee concludes that complete, or even heavy, reliance on third-party auditors may have the effect of contributing to a compliance mentality and be counterproductive to establishing a culture of safety. The comment period for the notice of proposed rulemaking was closed on November 14, 2011. Because the committee's report was not completed by that date and a final rule had not been issued as of the date of issuance of this report, the committee did not specifically address the proposed rule in detail in this report and did not make a formal comment to the proposed rule during the comment period.

REFERENCES

Abbreviations

API	American Petroleum Institute
BOEMRE	Bureau of Ocean Energy Management, Regulation, and Enforcement
NAE-NRC	National Academy of Engineering–National Research Council
NRC	National Research Council

API. 1993. *Recommended Practice for Development of a Safety and Environmental Management Program for Offshore Operations and Facilities,* 1st ed. API RP 75. API, Washington, D.C.

API. 2004. *Recommended Practice for Development of a Safety and Environmental Management Program for Offshore Operations and Facilities,* 3rd ed. API RP 75. API, Washington, D.C.

BOEMRE. 2010. Oil and Gas and Sulphur Operations in the Outer Continental Shelf—Safety and Environmental Management Systems. *Federal Register,* Vol. 75, No. 199, Fri., Oct. 15, 2010, pp. 63610–63654.

BOEMRE. 2011. Oil and Gas and Sulphur Operations in the Outer Continental Shelf—Revisions to Safety and Environmental Management Systems. *Federal Register*, Vol. 76, No. 178, Wed., Sept. 14, 2011, pp. 56683–56694.

NAE-NRC. 2011. *Macondo Well–Deepwater Horizon Blowout: Lessons for Improving Offshore Drilling Safety*. National Academies Press, Washington, D.C.

National Commission on the BP *Deepwater Horizon* Oil Spill and Offshore Drilling. 2011. *Deep Water: The Gulf Oil Disaster and the Future of Offshore Drilling*. Report to the President. http://www.oilspillcommission.gov/sites/default/files/documents/DEEPWATER_ReporttothePresident_FINAL.pdf.

NRC. 1990. *Alternatives for Inspecting Outer Continental Shelf Operations*. National Academies Press, Washington, D.C.

1

Introduction

The Outer Continental Shelf Lands Act provides for the jurisdiction of the United States over the submerged lands of the Outer Continental Shelf (OCS) and assigns the authority to lease such lands for certain purposes, such as mineral development, to the Secretary of the Interior. In 1982, after almost 30 years of divided agency responsibility in administering the Outer Continental Shelf Lands Act within the Department of the Interior (DOI), the secretary established the Minerals Management Service (MMS) from parts of the Bureau of Land Management and the U.S. Geological Survey to consolidate and carry out the department's authority for the nation's offshore oil and gas program.

In the aftermath of the *Deepwater Horizon* explosion, blowout, and oil spill in April 2010, DOI restructured MMS by transferring its revenue management functions to a new office and renaming the nonfiscal responsibilities of the agency the Bureau of Ocean Energy Management, Regulation, and Enforcement (BOEMRE).[1] On October 1, 2011, BOEMRE was further divided into two separate bureaus: the Bureau of Ocean Energy Management (BOEM) and the Bureau of Safety and Environmental Enforcement (BSEE). This report is most directly concerned with BSEE because of its delegated authority for safety and environmental oversight of OCS oil and gas operations, including permitting; inspections;

---

[1] On May 19, 2010, Secretary Salazar started the process of dividing MMS into three distinct parts through the issuance of Secretary Order 3299. On October 1, 2010, the royalty and revenue management functions of MMS, including, but not limited to, royalty and revenue collection, distribution, auditing and compliance, investigation and enforcement, and asset management for both onshore and offshore activities, were officially transferred to the new Office of Natural Resources Revenue.

enforcement of safety and environmental regulations; and oil spill response, training, and environmental compliance programs.[2]

BSEE's regulatory authority includes oversight responsibility with respect to the offshore platforms involved in drilling and production of oil and natural gas. Before November 2011, BSEE's oversight authority included the responsibility to conduct safety inspections of each platform at least annually as well as periodic unannounced "spot" inspections, the intent of which was to make offshore facilities safer. The belief was that the inspection process would encourage owners and operators to develop a healthy and viable safety culture on offshore facilities and that, if potential problems existed, they would be identified during the inspection process and subsequently addressed, thereby reducing the likelihood of a major incident.

STUDY CONTEXT

In 1990, the Committee on Alternatives for Inspection of Outer Continental Shelf Operations, under the auspices of the Marine Board, reviewed the MMS OCS inspection program and made several recommendations for improvement (NRC 1990). At that time, the inspection program mostly focused on facilities and whether they met certain standards. At each visit, inspectors worked through a checklist of potential incidents of noncompliance (PINCs). Among other determinations, the committee found the following:

1. The emphasis on compliance with hardware-oriented PINCs fostered an attitude of "compliance equals safety" that can actually "diminish the operator's recognition of his primary responsibility for safety" (NRC 1990, p. 80).

---

[2] In general, BOEM exercises the conventional (e.g., oil and gas) and renewable energy–related management functions of DOI and is responsible for the functions of DOI's offshore energy program related to leasing, environmental studies, National Environmental Policy Act analysis, resource evaluation, and economic analysis. BSEE oversees the safety and environmental enforcement functions of such programs including, but not limited to, the authority to inspect; investigate; summon witnesses and produce evidence; levy penalties; cancel or suspend activities; and oversee safety, response, and removal preparedness (http://www.boemre.gov/ooc/newweb/frequentlyasked questions/frequentlyaskedquestions.htm).

2. The "majority of accident events occurring on the OCS in a representative year (1982) were related to operational and maintenance procedures or human error that are not addressed directly by the hardware-oriented PINC list" (NRC 1990, p. 81).
3. "Third-party inspection by private sector contractors (alternative 4) would not diminish and would probably increase the tendency of operators to abdicate safety responsibility to the inspecting organization" (NRC 1990, p. 81).
4. "Self inspection (alternative 5), while it would pinpoint the operator's responsibility, would be unsuitable because the MMS oversight function would be too tenuous" (NRC 1990, p. 82).

The report recommended that inspections instead focus on a sample of PINCs and devote greater resources to unannounced inspections as well as increased analysis of incidents and accidents and data collected by inspectors. MMS should "place its primary emphasis on detection of potential accident-producing situations—particularly those involving human factors, operational procedures, and modifications of equipment and facilities" (NRC 1990, p. 83).

To make the detection of potential accident-producing situations more useful, the committee recommended that the quality and quantity of inspection data be considerably enhanced to allow MMS to take a more risk-assessment approach to inspections. Ultimately, the committee hoped that MMS would collect sufficient information about each platform to allow for development of risk indices that MMS could use to allocate more of its resources to platforms at higher risk. In the main, however, the committee stressed that the private operator was the primary agent responsible for ensuring safe operations and that MMS should structure its program to reinforce that awareness among operators.

MMS adopted some of the recommendations made in the 1990 report and spurred the offshore oil and gas industry to develop American Petroleum Institute (API) Recommended Practice (RP) 75, *Recommended Practice for Development of a Safety and Environmental Management Program for Offshore Operations and Facilities* (API 1993, 2004). This document recommends that the industry adopt management principles of planning, organizing, implementing, and measuring in managing safety in the same way that companies manage the remainder of their operations.

It includes specific guidance on elements required to carry out these management functions.

The industry was encouraged to adopt safety and environmental management programs voluntarily. In mid-2009, MMS proposed a rule that would have required offshore operators to adopt four of the 12 elements of API RP 75.

In April 2009, MMS again approached the Marine Board to request that the present study be conducted to review the MMS inspection program for offshore facilities to assess its effectiveness in protecting human safety and the environment. The committee was appointed in November 2009 and held its first meeting the following month. In March 2010, a subgroup of the committee made site visits to the MMS Pacific OCS Region and to the California State Lands Commission. The committee also scheduled a site visit in May of that year to the MMS Gulf of Mexico Region. The visit, however, was overtaken by the unfolding events and ensuing investigations of the *Deepwater Horizon* disaster (BOEMRE 2011b; NAE-NRC 2011; National Commission on the BP *Deepwater Horizon* Oil Spill and Offshore Drilling 2011; USCG 2011) and subsequent reorganization of MMS into the Office of Natural Resources Revenue and BOEMRE. During this process, agency officials asked that this project be put on hold while the agency reevaluated its approach to safety.

In October 2010, BOEMRE issued a final rule requiring adoption of API RP 75 with minor revisions as defined in the rule and retitled "Oil and Gas and Sulphur Operations in the Outer Continental Shelf—Safety and Environmental Management Systems" (SEMS) (BOEMRE 2010). The SEMS rule became effective on November 15, 2011 (30 CFR 250, Subpart S). It lays out multiple requirements for safe and environmental operations, including requiring specific written plans for operating practices, hazards analysis, management of change, safe work practices, training, mechanical integrity, emergency response, and incident reporting. API RP 75 recommends that practices be audited by a qualified party, which could include individuals employed by the same company, on a regular schedule. As stated in the rulemaking,

> The ultimate goal of SEMS is to promote safety and environmental protection during OCS activities. The protection of human life and the environment are the top priorities and objectives of this rule. While it is

difficult to provide absolute quantification of the benefits of the lives saved and risks avoided due to this regulation, the BOEMRE believes that implementation of a comprehensive SEMS program will avoid accidents that could result in injuries, fatalities, and serious environmental damage based upon BOEMRE's incident analysis. In addition, an increase in a system's level of safety leads to reduced material losses and enhanced productivity. (BOEMRE 2010, p. 63644)

In the SEMS rule, BOEMRE recognized that its inspection program was too focused on mechanical failures and that such failures represent a small minority of incidents. With issuance of the final rule, BOEMRE's approach to safety and environmental protection shifted from reliance solely on inspections of hardware-oriented PINC items to also requiring operators to specify how they will manage safety holistically to avoid injury and spills.

After the SEMS rule, BOEMRE officials recognized that additional provisions were needed; thus, they issued a notice of proposed rulemaking, "Oil and Gas and Sulphur Operations in the Outer Continental Shelf—Revisions to Safety and Environmental Management Systems," referred to as "SEMS II," on September 14, 2011 (BOEMRE 2011a).[3] The revisions in the proposed rule pertain to

- Developing and implementing stop work authority and ultimate work authority,
- Requiring employee participation in the development and implementation of SEMS programs,
- Establishing requirements for reporting unsafe working conditions,
- Requiring independent third parties to conduct audits of operators' SEMS programs, and
- Establishing further requirements relating to conducting job safety analysis for activities identified in an operator's SEMS program.

Because SEMS II has not yet been adopted and is subject to modification, the committee did not specifically evaluate the audit requirements for each of these issues in this study.

---

[3] See also http://www.gpo.gov/fdsys/pkg/FR-2011-09-14/pdf/2011-23537.pdf#page=1.

STUDY OBJECTIVE AND CHARGE

In late 2010, following restructuring of MMS, BOEMRE requested that the scope of the committee's study be changed from a review of the agency's previous offshore platform safety and environmental inspection program to one that provided guidance on how the agency should evaluate and ensure the effectiveness of the implementation of the new SEMS practices that were required of offshore operators as of November 15, 2011. As a result, this project was refocused, and the committee resumed its work in late January 2011.

Under the new agreement with BOEMRE, the committee was renamed the Committee on the Effectiveness of Safety and Environmental Management Systems for Outer Continental Shelf Oil and Gas Operations, and its charge was revised. The following charge, as modified in late January 2011, was presented to the committee:

> This project will recommend a method for assessing the effectiveness of an operator's Safety and Environmental Management System (SEMS) on any given offshore drilling or production facility. In addition, the committee will prepare a brief interim report in April 2011 that will provide a listing of potential methods for assessing effectiveness along with the pros and cons of each method as they are known to that point. The committee will address methods to maximize the implementation effectiveness of individual SEMS rather than the adequacy of the Final Rule of October 2010 requiring SEMS to mitigate safety and environmental risk of offshore platform operations.
>
> The committee's assessment of effective methods will focus on the safety and environmental risks of offshore production until after the release of the report of the NAE/NRC Committee for the Analysis of Causes of the *Deepwater Horizon* Explosion, Fire, and Oil Spill to Identify Measures to Prevent Similar Accidents in the Future, which is expected in June 2011 [but was actually released in December 2011]. The committee's assessment of effective methods for safety and environmental risks of drilling will take into account the findings and recommendations of the NAE/NRC committee.

The interim report was released in June 2011. The present final report, which was developed through open- and closed-session meetings, presentations, discussions, and subsequent correspondence, presents an assessment of different methods for assuring the adequacy of offshore operators' SEMS programs and recommends what it considers to be the best approach. The report also takes into consideration the findings and recommendations of the National Academy of Engineering–National Research Council Committee for the Analysis of Causes of the *Deepwater Horizon* Explosion, Fire, and Oil Spill to Identify Measures to Prevent Similar Accidents in the Future, which released its final report on December 14, 2011 (NAE-NRC 2011).

ORGANIZATION OF THE REPORT

Chapter 2 presents an assessment of the role of SEMS, its goals, and its potential impact on an operator's culture of safety. Chapter 3 contains a description of nine different methods that could be used to evaluate the effectiveness of an operator's SEMS program and discusses some of the advantages and disadvantages of these methods. Chapter 4 presents currently used approaches for assessing safety management in other regulatory agencies in the United States, as well as in the offshore oil and gas industry in a few other countries that have a charge similar to that of BSEE. The chapter also includes a brief description of the potential role of the newly created Center for Offshore Safety. Chapter 5 discusses the role of BSEE in evaluating SEMS programs, including the use of inspections and audits, the training and qualifications of auditors, audit criteria and procedures, and the competence of inspectors and auditors in ensuring effectiveness. Chapter 6 presents the committee's conclusions and recommended approach.

2

Role of Safety and Environmental Management Systems in Establishing a Culture of Safety

From the most literal (and simplistic) perspective, the Committee on the Effectiveness of Safety and Environmental Management Systems for Outer Continental Shelf Oil and Gas Operations (the committee) could have achieved its goal by first reviewing the documented requirements of a Safety and Environmental Management Systems (SEMS) program and then describing methods for determining whether those specified elements were being used. For example, the committee could have determined ways of assessing whether a hazards analysis was in place (e.g., by creating a checklist or defining a process) and then identified ways to document evidence that the results of the hazards analysis were being addressed. Such an approach would have resulted in recommendations for auditing compliance to a defined standard (e.g., the requirements of SEMS). That defined standard would, in practice, become the minimum standard.

The National Commission on the BP *Deepwater Horizon* Oil Spill and Offshore Drilling (2011) observed:

> The record shows that without effective government oversight, the offshore oil and gas industry will not adequately reduce the risk of accidents, nor prepare effectively to respond in emergencies. However, government oversight, alone, cannot reduce those risks to the full extent possible. Government oversight must be accompanied by the oil and gas industry's internal reinvention: sweeping reforms that accomplish no less than a fundamental transformation of its **safety culture**. (p. 217, emphasis added)

The committee agrees with the presidential commission that a transformation of the industry's safety culture is necessary and believes that an approach based on compliance with a minimum standard will not achieve that goal. In fact, the committee believes that overemphasis on compliance with a minimum standard can actually work against that intended objective.

An effective SEMS program is a necessary and critical component of offshore safety. Without a well-reasoned, well-documented method of coordinating action, consistently safe operations are simply not possible. Nevertheless, as important as a SEMS program is, it alone cannot ensure that the people actually doing the work (whether planning or designing onshore or working offshore) make the choices and take the actions necessary to ensure safety. Safe and effective operations are, in part, indicative of an effective safety management system (SMS); however, safe and effective operations are not created solely by the management system, but by a set of diverse components. Factors such as a culture of blame and a lack of mindfulness of risk, organizational commitment, and trust have been shown time and again to be contributors to high-profile tragedies in the petroleum industry and elsewhere (DNV 2011; Hopkins 2004, 2006). Because a SEMS program cannot reliably control what people choose to do on the job, the mere existence of a documented SEMS plan is not sufficient to ensure prevention of major accidents.

The spirit of SEMS, whether as defined in American Petroleum Institute Recommended Practice 75 (API 2004) or in other similar approaches, is not intended to be strictly a paper exercise. The way that SEMS is actually implemented, even by different divisions in the same organization, can produce different results. By way of example, airlines use the very same equipment under similar conditions and have very similar written maintenance and operational processes and procedures, but differences in passenger risk of some 40 times have been documented (PSA Norway 2002; Reason 1997). Getting the people who actually do the work to make the right choice, every time, even when they are outdoors in the cold rain, under tight time constraints, and when no one is looking is different from having an auditable SEMS program in place; people have called these differences in terms of the way organizations operate "organizational culture."

WILL SEMS PROMOTE A CULTURE OF SAFETY?

Although a culture of safety is a goal of many organizations and attempts are made to measure it, people often find describing a safe culture in concrete terms difficult. According to James Reason, a definition of culture captures most of its essentials: "Shared values (what is important) and beliefs (how things work) that interact with an organization's structures and control systems to produce behavioural norms (the way we do things around here)" (Reason 1983, p. 294, and 1997, p. 192). According to Booth, the United Kingdom Health and Safety Commission defined safety culture in the following way:

> The safety culture of an organization is the product of individual and group values, attitudes, competencies, and patterns of behaviour that determine the commitment to, and the style and proficiency of, an organization's health and safety programmes. Organizations with a positive safety culture are characterized by communications founded on mutual trust, by shared perceptions of the importance of safety, and by confidence in the efficacy of preventive measure. (Booth 1993, p. 5)

Culture is critical in the choices people make and can promote or inhibit safe choices. Many people, according to Reason (1997, p. 192) believe that "a safety culture can only be achieved through some awesome transformation," such as might occur as a result of a catastrophic organizational accident. He believes, however, that these changes are often short-lived because a safety culture is not something that springs up ready-made from the organizational equivalent of a near-death experience, but, in fact, "emerges gradually from the persistent and successful application of practical and down-to-earth measures" (Reason 1997, p. 192).

As major incident investigations have shown (e.g., Borthwick 2010; BP U.S. Refineries Independent Safety Review Panel 2007; CAIB 2003; CSB 2007; Cullen 1990; National Commission on the BP *Deepwater Horizon* Oil Spill and Offshore Drilling 2011), the existence of an effective safety culture is fundamental to the creation of a safe work environment. In the incidents cited here, and many others, the lack of a positive safety culture has been cited as a major contributor. It is, therefore, a logical supposition that safe operation in a high-hazard industry requires an effective culture of safety. The term "safety culture" is often misconstrued as indicating a means of convincing individuals to comply with regulations and procedures; the term is more effective, however, when viewed as the intrinsic value of the importance of safety (HSE 2011).

Several industries and regulatory bodies in the United States as well as other countries have policies and guidelines for creating a positive culture of safety. The U.S. Nuclear Regulatory Commission (U.S. NRC) created a policy outlining its expectation that individuals and organizations performing regulated activities establish and maintain a positive safety culture commensurate with the safety and security significance of their activities and the nature and complexity of their organizations and functions. U.S. NRC outlined several traits that are common in an effective culture of safety. These are cited in the report *Macondo Well—Deepwater Horizon Blowout: Lessons for Improving Offshore Drilling Safety* (NAE-NRC 2011, pp. 92–93) and are adapted here with additional information from Reason (1997) and HSE (2011):

- *Leadership safety values and actions.* Genuine values are consistently communicated by leadership through visible commitment to safety; values and actions are not tied to leadership's personality or to commercial concerns. Leadership's commitment demonstrates a high level of concern for safety throughout the organization through resource allocation and priority support for safety versus production. Organizational leaders also visibly influence and lead by demonstrating their values through their decisions and actions, thereby ensuring that employees see that the commitment to safety is genuine.
- *Problem identification and resolution.* Issues are identified, evaluated, addressed, and corrected promptly.
- *Personal accountability.* Personal responsibility for safety is accepted by each individual. Workers take a proactive role and ownership in their own safety and that of colleagues.
- *Work processes.* Planning and control of work processes is implemented to maintain safety.
- *Continuous learning.* The organization works as a learning organization—that is, an organization that pursues current knowledge and collects data and information to become and remain informed and that adapts as this new knowledge and information are gained.
- *Environment for raising concerns.* The organization maintains a safety-conscious work environment in which personnel feel free to raise safety concerns without fear of retaliation, intimidation, harassment, or discrimination. Reason (1997) describes this type of environment as a willing reporting culture, in which decisions and changes necessary for success are made following investigations.

- *Effective safety communication.* Communications within the organization maintain a focus on safety to ensure that mixed messages for competing priorities are not the norm. Knowledge and experience are shared across organizational boundaries. This sharing can be especially important when different companies are involved in various phases of the same project. Knowledge and experience are also shared vertically within the organization.
- *Respectful work environment.* Trust and respect permeate the organization. The workforce is treated with dignity and respect.
- *Questioning attitude.* Individuals avoid complacency and continuously challenge existing conditions and activities to identify discrepancies that might result in unsafe conditions. No worker hesitates, at any time, to question work practices at any level, and this questioning is considered part of everyday work conversations. As noted by Meshkati (1999), a facility that emphasizes and fosters a culture of safety encourages employees to develop a questioning attitude and a rigorous and prudent approach to all aspects of their jobs and to establish open communication between line workers and middle and upper management.

According to Reason (1997, p. 196), a safety culture has four critical subcomponents:

- *A reporting culture:* People are willing to report their own errors and near misses.
- *A just culture:* Individuals are encouraged when they provide essential safety-related information.
- *A flexible culture:* Control changes according to the expertise needed in specific situations because there is respect for members of the workforce who have the skills, experience, and abilities to respond to the situation.
- *A learning culture:* The organization and the workforce learn and make changes as needed.

These four subcomponents interact to create an *informed* (i.e., safe) *culture* that will reduce the likelihood of organizational accidents.

Another way of thinking about safety culture is that, in a safety culture, the subjective aspects of the organization (attitudes, perceptions, and values) are integrated with objective processes and systems. It is this integration and collaboration that support effective safety performance.

Role of SEMS in Establishing a Culture of Safety

	Able to How: Process	Want to Why: Purpose
Organization	**Mechanism** What do people read or write . . .?	**Culture** Why do people . . . if it wasn't in their immediate interest?
Individual	**Competency** How are individuals capable of . . .?	**Motivation** Why would a totally selfish person . . .?

FIGURE 2-1 Interaction of culture and process.

One useful way to explain the interaction between process and culture is with the matrix in Figure 2-1. This matrix illustrates the elements required for an action to occur reliably in a real organization. For something to occur reliably, the *organization* as a whole and each *individual* in the organization need to be *able to* accomplish the action and need to *want to* do so. The *organization–able-to* quadrant of the matrix describes the mechanism an organization would use to operate safely. The SEMS plan and supporting documentation correspond to the *organization–able-to* quadrant.

Without an effective SEMS (or SEMS-like) plan and appropriate documentation, it is very unlikely that an organization could operate safely; however, great plans and supporting documentation do not mean the organization will be safe. The *individual–able-to* quadrant of the matrix is competency; it describes how people as individuals are capable of executing the requirements of safe operations. There may be great plans, but without competent individuals, they cannot be carried out.

The *individual–want-to* quadrant is motivation; it describes those factors in the organization that would cause a totally self-interested person to want to work safely. For example, if people really are totally unmotivated to report incidents (e.g., because bonuses are lost or because the paperwork is just too much of a hassle) then more training on how to spot incidents will not address the issue. The individual must be motivated and empowered to work safely.

Finally, the *organization–want-to* quadrant is the culture or behavioral norms that cause people to act properly even when no one is looking and when it is not in their immediate best interest. A healthy safety culture causes people to report events accurately, even when they are at fault, because truthfulness is the norm.

If one of these elements is missing, there will be a bottleneck in the organization's ability to work safely and with environmental responsibility, and more emphasis on the other elements will not address the problem. If either motivation or culture is missing, lack of additional training or lack of more detailed processes will probably not be the root cause of an incident. The true root cause will probably be something missing in the organization's culture or the individual's motivation.

To build a culture of safety from an organizational level there must be

- Mechanisms that establish structure and control by specifying what is needed for safe operation and providing for checking to see that these specifications are being followed (SEMS' organizational element), and
- Actions that establish safety norms by encouraging people to act properly even when no one is looking or when it is not in their immediate best interest.

To build a culture of safety from the individual's level there must be

- Mechanisms that build competency by developing individual knowledge and skill (SEMS' requirements for training, operating procedures, and safe work practices), and
- Actions that build the motivation of a totally self-interested person to act in accordance with behavioral norms.

An organization's culture is created by thousands of individual actions and by leadership at all levels; but the culture must be owned by the top leadership, in addition to the middle managers and the line workers, because "[n]o matter what regulatory system is used, safe operations ultimately depend on the commitment to systems safety by the people involved at all levels within the organization" (NAE-NRC 2011, p. 116).

According to Peters and Waterman (1982), if there is a strong culture, all levels of the organization will have shared goals and values. The culture of safety cannot be built or sustained through publishing statements from the chief executive officer and human resources department, posting notices in company internal and external communications, punishing individuals for incidents of noncompliance (INCs), rewarding individuals for a lack of INCs, or reading perfunctory safety minutes prior to meetings. It is something that the leadership must live. The management of safety within an organization is ultimately a reflection of its safety culture. A poorly designed and implemented SEMS program can work against creating the conditions needed for a healthy safety culture to develop. Conversely, effective implementation of a SEMS program is expected to have a positive impact on the safety culture of companies operating on the U.S. Outer Continental Shelf; however, whether it will do so will not be known until trend data are available and analyzed.

To exist and grow, a culture of safety requires reciprocity between corporate management and individual employees' values, beliefs, and perceptions. A SEMS program can create the backbone of the safety culture upon which organizations build these internal reciprocal relationships. A culture of safety requires commitment, engagement, and execution from all levels of the organization. It is this ownership and engagement that reshapes safety culture into a continuing, long-term commitment to improve. The committee agrees with the NAE-NRC committee that

> SEMS will require companies to adopt both a *top-down* and a *bottom-up* safety culture. Safe . . . operations cannot be achieved solely through regulations, inspections, or mandates. They will only be realized when there is a full commitment to system safety, from the board room to the rig floor, and through recognition that a focus only on occupational safety will not ensure system safety. Compliance with either prescriptive regulations or standards related to achieving specific safety goals need[s] to be considered a minimum requirement and not necessarily a way to meet duty of care obligations." (NAE-NRC 2011, pp. 119–120)[1]

A common problem for some companies is the tension between organizational mandates regarding safety and pressure for efficiency in terms of time and money. Companies continually make decisions that trade safety off against other objectives (e.g., time and cost). Without a framework that keeps safety concerns elevated to an appropriate level,

---

[1] The reader is referred to Chapter 5, "Industry Management of Offshore Drilling," of the NAE-NRC (2011) report for additional information about system safety, safety culture, and high-reliability organizations. This information is not strictly limited to offshore drilling operations, but is applicable to offshore oil and gas facilities in general.

inefficient, even disastrous, decisions will ultimately be made. This can happen when the conflict of responsibility and accountability with respect to many different organizational goals (e.g., safety, time, and production) ensures that the target with the most forceful message from top management will prevail. Building trust that top management will support safety decisions made by personnel throughout the organization, even when they are in conflict with other priorities, is the only way to achieve a culture of safety. SEMS alone cannot build this trust.

To achieve reliably safe operations, more than a well-defined SEMS program is needed. People in the organization must actually use the SEMS program and improve its implementation on a continuing basis. Thus, auditing of SEMS programs should extend beyond verifying the existence of a SEMS program—and the existence of documentation that supports its use—to assuring that what is described in the SEMS plan is actually the way people in the organization think and work.[2] Effective measurement of the efficacy of a SEMS program must extend beyond verifying the paper records of the program to examining how the SEMS plan is used to guide what individuals in the organization do to ensure safe and environmentally responsible operations.

GUIDING QUESTIONS FOR EVALUATION OR AUDIT

Any audit process offers multiple opportunities for checking the strength and effectiveness of each platform's realization of SEMS. A sequence of guiding questions provides a preliminary structure for the audit:

1. *Is a SEMS plan in place?* Is the plan complete? Is there a document to read? Has the owner or operator structured a plan that covers all the necessary personnel, equipment, and situations?
2. *Is the plan feasible and effective?* Given that a plan is in place, how good is the plan at reducing risks? If the steps outlined in the plan

---

[2] Individual, organizational, and technical factors and their impact on the culture of safety are all considered in the various philosophies, frameworks, and techniques espoused by leading researchers who study highly complex systems, high-reliability organizations, and the like. For more detailed discussions of this issue, the reader is referred to the following sources (to name but a few): ABS (2012), Bea (2002), Hopkins (2004, 2006), LaPorte and Consolini (1991), Reason (1997), Scarlett et al. (2011), Schein (1992, 2004), Weick (1987), and Weick and Sutcliffe (2001, 2007).

are followed, will they be successful in meeting program safety goals? Are sufficient resources available to comply with the plan? How does the plan compare with plans that have been developed for other similar platforms and have been shown to be effective?
3. *Do personnel know about the plan?* A well-written and carefully thought-out program will not succeed if the personnel required to follow it are not aware of it. Is there a way to track components of SEMS with the necessary personnel? As personnel are replaced, is there a process by which new personnel are introduced to their responsibilities? Is the plan pervasive throughout the organization?
4. *Can and do personnel effectively carry out the plan?* That personnel are aware of the program does not mean that they can follow it effectively. Is a training program in place? Are there periodic tests and drills with which personnel can demonstrate their familiarity and expertise with details of the plan?
5. *Is the plan affecting safety?* The goals of SEMS programs are to improve both occupational and process safety. Are metrics that permit verification of the SEMS plan being recorded and tracked? Is the plan being used to instill and encourage a healthy safety culture? Long-term effectiveness can only be assessed through the comparison of tracked measures with baseline data. Are near-miss events related to occupational and process safety being recorded and evaluated? A careful definition of performance metrics would allow for comparisons across platforms, rigs, operations, lessees and operators, and regions. It would also facilitate international comparisons.

Each question requires a different audit approach; a different data collection requirement; a different audit schedule; and, potentially, a different type of trained auditor. Strengths and weaknesses of alternatives for these options are discussed in the following sections.

ASSESSING THE EFFECTIVENESS OF SEMS AND ITS EFFECT ON CULTURE

With its inspection and audit programs, the Bureau of Safety and Environmental Enforcement (BSEE) is in a unique position to influence how SEMS is implemented and integrated into an organization. As discussed above, more than a well-defined SEMS program is needed to achieve reliably safe operations; people in the organization must actually implement the program and improve it on a continuing basis. An effective audit program would extend assurance beyond verifying paper records to investigating how the program is used to guide what individuals in the organization do to ensure safe and environmentally responsible operations.

For example, issuing INCs for failure to comply with prescriptive regulations leads to an attitude that compliance equals safety and does not influence behavior beyond the minimum standard. Because tacit knowledge exceeds explicit knowledge by several times, it is not possible to define a set of rules that, if followed exhaustively, will create safety. People need to understand the objectives and work toward those objectives, not blindly follow a minimum standard.

Even worse, issuing INCs as punishment after the fact for inappropriate behavior (the stick half of a carrot-and-stick approach) can create a culture of fear and blame. Practical experience in everything from child raising to conforming to a group norm has shown that fear of punishment can be used to provide a minimum level of expected behavior, but fear of punishment does not normally affect basic attitudes.

More will be described later in this report, but briefly, BSEE has a critical role in

- Auditing for the existence of a SEMS program and for its built-in improvement mechanisms and
- Grading and counseling before the fact to help management establish norms and motivation (the carrot).

Grading and counseling will help corporate leadership better understand how to strengthen the actual structure, controls, and competency that exist in its operations. BSEE can also help corporate leadership understand how to improve the actual state of behavioral norms and motivation in its operations.

Such an evaluation system should not be strictly objective or quantitative and cannot be a matter of pass or fail. The evaluation system will need elements such as interviews with a sample of workers and first-level supervisors, grading of each of the elements of SEMS, and reviews of results with leadership. This process must be repeated periodically to find trends, and evaluation results should be publicly reported to provide both a carrot and a stick. Most importantly, it will require changing from an INC mentality (punishment) to a cooperative mentality (consultation and advice).

A WORD OF HOPE

Since 1968, the oil and gas industry has reduced lost-time incidents by some 97.5 percent (Figure 2-2), despite a large increase in hours worked. This change did not happen randomly. The industry has specifically focused on significantly improved occupational safety over the past few decades. Accomplishing this improvement required not only new processes (such as job safety analysis), but also cultural change. In the early 1970s, operations people actually quipped, "If you aren't missing a finger, it means you haven't worked very hard." No one says this today, and if someone were to say it, he or she would be viewed by many of his or her peers with disdain.

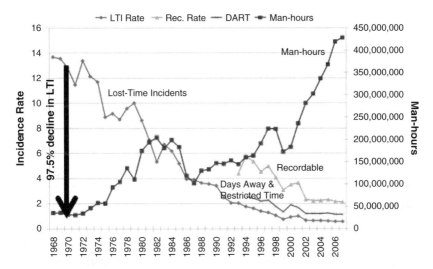

FIGURE 2-2 Industry safety metrics. For 2007, man-hours are estimated and third-quarter incidence rates are used. (LTI = lost-time incidents; Rec. = recordable; DART = days away and restricted time. SOURCE: IADC 2011.)

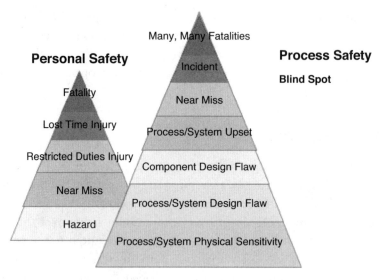

FIGURE 2-3 Personal versus process safety pyramids.
(SOURCE: Hopkins 2009.)

The focus on occupational (personal) safety has led to dramatic reductions in lost-time incidents, recordable incidents, and the like. However, organizations with a good occupational safety record are not necessarily managing large-scale risks—that is, system safety or process safety—appropriately, as illustrated in the Macondo well–*Deepwater Horizon* catastrophe (see NAE-NRC 2011). Managing process safety means ensuring "that safety is built into a system with the objective of preventing or significantly reducing the likelihood of a potential accident" (NAE-NRC 2011, p. 91) in order to manage the very rare but very high-consequence incidents that can lead to multiple losses of life, substantial property loss, and extensive environmental damage.[3] Figure 2-3 shows the difference between the occupational (personal) and process safety pyramids.

In the past, regulators and the industry have not focused as much on total system safety (which includes process safety) as they should. The committee believes that, with a properly constructed SEMS program

---

[3] For additional discussions of system safety, see for example, Leveson (2011), Rasmussen (1997), and Rasmussen and Svedung (2000).

that encompasses a clear focus and intentional action, the industry can improve process safety without compromising occupational safety.

In a widely circulated video, Brian Appleton, technical adviser to the Lord Cullen inquiry team into the *Piper Alpha* accident in the North Sea, makes the point that a safety audit that does not find deficiencies in an SMS should be suspect: "In safety, no news is not good news" (Appleton 1995). The committee heard similar sentiments in meetings with the California State Lands Commission and Petroleum Safety Authority Norway, two organizations that have extensive experience auditing SMS programs. That is, a pass–fail, INC-based audit of a SEMS program that does not find deficiencies is probably not a good audit. Such an audit will have a tendency to focus on written policies and procedures to determine whether they contain the exact wording required by 30 CFR 250, Subpart S, and operators will expend great effort to assure that the words are "correct" and the proper documentation is on file.

If BSEE's goal is, as it should be, to encourage a culture of safety so that individuals know the safety aspects of their actions and are motivated to think about safety, then the agency will need to evolve an evaluation system for SEMS that emphasizes the evaluation of attitudes and actions rather than documentation and paperwork. All of the elements of SEMS must be addressed, but it is more important that those who are actually doing the work understand and practice these elements than that these elements are documented.

Lord Cullen said of the *Piper Alpha* "permit to work" system, "The operating staff had no commitment to working to the written procedure; and . . . the procedure was knowingly and flagrantly disregarded" (Appleton 1995). An evaluation system that emphasized documentation may have missed the lack of a proper culture of safety on the *Piper Alpha*.

The remainder of this report contains the committee's justification and recommendations for how BSEE can assess the effectiveness of an operator's SEMS program while simultaneously promoting development of a fundamental transformation of the industry's safety culture. The report describes an approach that the committee believes will guide BSEE in playing a critical role in helping the industry transform its safety culture, with the goal of making the risk of working offshore as low as reasonably practicable (ALARP).

3

Methods for Assessing Effectiveness

Control is a vital management function by which operations are brought into compliance with predetermined standards that are established on the basis of planning and implementing systems to achieve the goals of an organization. It is axiomatic that to control, one must first measure. To measure, one must know the characteristics of the parameters on which measurements are being made. If measurements are to be made reliably, the influences that affect the measurement must be known. Operational results, causes, and effort can be measured. The data so acquired must be evaluated as to the impact on performance, which is a measure of the effectiveness of the actions. Decisions about effectiveness, therefore, are quite complex, in that they involve judgments about assessment, methods, and evaluation of data from operations. The degree of complexity increases with the complexity of the system being evaluated.

The Safety and Environmental Management Systems (SEMS) regulations require operators to develop and submit a SEMS plan to the Bureau of Safety and Environmental Enforcement (BSEE). Assessment of the effectiveness of an operator's SEMS program is an essential step toward improving the quality of SEMS application in practice. SEMS regulations prescribe specific audit requirements: a comprehensive audit 2 years from the initial implementation of the SEMS program and at least once every 3 years thereafter.

POTENTIAL ASSESSMENT METHODS

The breadth and depth of SEMS require that several methods be used to assess its effectiveness on an ongoing basis for continuous improvement in development and implementation. Operators, who are responsible for the development of a SEMS program, must develop a plan for assessing the implementation and performance of the program at the same time. The Committee on the Effectiveness of Safety and Environmental Management Systems for Outer Continental Shelf Oil and Gas Operations (the committee) has identified nine methods that may be used to assess the effectiveness of an operator's SEMS program:

1. Compliance inspections,
2. Audits,
3. Peer reviews and peer assists,
4. Key performance indicators,
5. Whistleblower programs,
6. Periodic lessee reports,
7. Tabletop exercises or drills,
8. Monitoring sensors, and
9. Calculation of risk with SEMS in place.

Some of these methods can be further subdivided. These nine methods are not mutually exclusive, and elements of each could be combined to develop the most effective evaluation program for a given operator. Table 3-1 summarizes the nine methods, which are discussed below, and notes pros and cons for each one.

Compliance Inspections

Compliance inspection is one of the simplest forms of SEMS verification. The intent is to verify, with little time and minimal inspector training, that at least portions of the SEMS program are operating. The compliance inspection is not meant to be a comprehensive audit such as that described below; rather, it provides a general indication of the state of the SEMS program by verifying specific components. Checklists may be used to conduct compliance inspections to ensure that documentation is compliant with the regulations. For example, the inspector may use a brief checklist to verify that SEMS items such as training (certificates) and operating procedures and emergency response plans are in place and that staff are familiar with the use of the latter two. Carefully crafted interviews of operational personnel can be very effective in determining whether workers understand how and why their actions lead to safer

TABLE 3-1 Summary of Methods for Assessing the Effectiveness of SEMS Programs

Method	Description	Pros	Cons	Notes
1. Compliance inspection	Onboard SEMS check by day-to-day BSEE inspectors; regional inspectors can also perform SEMS check	Maintains minimal compliance Provides regulatory presence at the operations level	Scope of SEMS check limited because of responsibilities for inspections of all other mandatory requirements	
a. Checklist	Checklist to ensure SEMS is in place on platform Checklist scope and details may vary	Simple to implement with minimal training May quickly identify deficiencies with SEMS program and implementation	May only assess compliance with paperwork or system; limited assessment of effectiveness of the SEMS program Platform specific; not a corporate-wide check Content and quality can vary extensively Must develop checklists	
b. Interviews, witnessing, and so forth	Interviews or other communication with platform personnel to determine whether they understand the SEMS program, including possible test drills May be concurrent with administering checklists	Can provide information to assess whether platform personnel are knowledgeable and use SEMS	Can be subjective Reliant on interviewer skills Additional SEMS training required, perhaps substantial Time consuming	California State Lands Commission program is an example

	Description	Advantages	Disadvantages	Notes
2. Audit	Review of implementation and quality of SEMS at both corporate and platform level. Platform level may be all platforms or a sampling. Scope (e.g., comprehensive or selected components) and details (time interval, auditing protocols) can vary	Proven method. Established auditing protocols available for process safety management (e.g., API, American Institute of Chemical Engineers). Scope and details can vary	Can only provide a reasonable assurance that the system is effective. Specific protocols need to be developed for defined scope. Auditor required to be expert at SEMS. Several auditors may be required in order to look at all SEMS areas	
a. Periodic audit	Planned in advance on a regular basis, typically 2- to 3-year intervals	Can be scheduled to meet BSEE requirements. Can be a comprehensive audit	Cost and time. Need to develop specific protocols for SEMS audit	Guidelines for meeting BSEE audit requirements
b. Surprise or random audit	Unannounced; a combination of randomly selected SEMS across all owners	Instantaneous assessment of state of SEMS implementation	May disrupt normal activities (e.g., drilling or testing). May not be comprehensive	"Surprise" means several days' notice, not instantaneously
c. Event-driven audit	Triggered by events such as injury or death, pollution, a near miss, and noncompliance	Immediately corrects SEMS issues, if applicable	Reactive, lagging assessment. May not reflect processes in place prior to incident	May be required in any case by regulations
3. Peer review, peer assist	Assessment of SEMS implementation by a team composed of peers from the industry	Team is qualified and experienced in SEMS. Nonthreatening identification of catastrophic weaknesses and opportunities to improve. Good potential to learn from each others' SEMS	Independence may be questioned. Potential conflicts of interest and confidentiality. Potential legal liability issues related to discoverability of recommendations and recommendations given in good faith that have poor outcomes	

(continued on next page)

TABLE 3-1 (*continued*) Summary of Methods for Assessing the Effectiveness of SEMS Programs

Method	Description	Pros	Cons	Notes
4. Key performance indicators	Use metrics from corporate- or platform-specific data to assess SEMS effectiveness Metrics can be currently reported ones (e.g., INCs, spills, accidents, near misses) or expressly developed new ones [e.g., number of changes (i.e., MOC), SEMS INCs]	Quantitative Easy to implement Can be automated and reported to BSEE regularly (quarterly) Could be used to identify specific problem platforms BSEE databases available for analysis	Unclear as to how current metrics relate to SEMS effectiveness New metrics may need to be developed If metrics do not accurately reflect safe conditions, they could create complacency	BSEE can establish specific SEMS INCs
5. Whistleblower program	Owner's policy and programs for anonymous reporting of events or situations by employees or other persons to complement normal reporting and communication channels that would lead to better SEMS implementation	Proactive for identifying corrective actions Evidence of management's commitment to SEMS Engages staff day to day Easy to implement	Lagging indicator of problems already in place Disgruntled persons can report false information Dependent on culture Requires follow-up program and fast and transparent follow-up by owner	May be available in other industries (e.g., nuclear, aviation)
6. Periodic lessee report	Quarterly, biannual, or yearly specific report from the lessee on the status and effectiveness of its SEMS program Scope and details of these voluntary reports can vary	Keeps SEMS relevant and recent in terms of operator's processes As voluntary submissions, these may be useful when performing mandatory SEMS audits	Accuracy of self-report can be questioned Can be onerous on operator Scope and detail are not defined and may need to be developed	Report context and content are current and relevant; may be corporate level rather than platform specific

7. Tabletop exercise or drill	Planned or surprise drill with specific actions to test SEMS; similar to spill drills Can vary from simple to complex exercises, depending on the scope of SEMS tested	Can become a subset of existing drills True reflection of SEMS in action	Cannot test all SEMS—would have to be a selection Would require much preplanning by owner and BSEE Can only be applied to a limited number of facilities Time consuming May require dedicated BSEE personnel and skill set
8. Monitoring sensors	Tracking onboard sensors to establish specific metrics for SEMS purposes	Quantitative SEMS measure Possible future development of SEMS-specific sensors Can send data back to shore for evaluation	Need to identify how these sensors may reflect SEMS issues
9. Calculation of risk with SEMS in place (QRA)	Specific quantitative methods that use owner's SEMS program as well as statistics from platform operations to determine effectiveness of SEMS over time	Measurable Can see changes in performance over time	Quantitative, results can vary between QRA approaches Need data over time to see trends Need baseline data for statistical analysis Output depends on model assumptions and details

NOTE: API = American Petroleum Institute; INC = incident of noncompliance; MOC = management of change; QRA = quantitative risk assessment.

operations and can lead to an understanding of the underlying safety and environmental culture of the organization. These types of interviews are also part of normal audit procedures.

Audits

An audit of a SEMS program should be a classic audit that consists of a comprehensive, systematic collection and review of information to ensure the program is being maintained and operated as intended. Where possible, the audit should verify objective evidence that shows conformance with the SEMS program. The audit can be performed by one or more internal staff (a first-party audit), by an associated outside organization (a second-party audit), or by a completely independent organization (a third-party audit). Audits may be periodic, surprise or random, or event driven. Event-driven audits are particularly effective in leading to an understanding of what went wrong and why and are often the impetus for major changes in industry approaches and regulatory oversight. The current BSEE SEMS regulation that went into effect November 14, 2011, allows first-, second-, and third-party audits, but the pending SEMS II regulation, as proposed in the September 2011 notice of proposed rulemaking (BOEMRE 2011a), authorizes only independent third-party audits. Complete or partial audits of an operator's SEMS program could be conducted, as justified by reports from inspectors, reviews of operators' audit reports, incidents, or events.

Peer Review and Peer Assist

Often simply referred to as "peer assist," this method of assessing effectiveness engages several respected industry peers from outside the organization, including other operators, in reviewing the company's compliance performance and SEMS implementation. The reviewers then suggest helpful ideas for improvement. There may or may not be formal documentation.

Peer assists are a common intracompany and intercompany activity for technical and economic issues and have been found to work well in other contexts. There are different protocols for this method (e.g., different levels of required response to peer recommendations). For example, a peer assist can be

- An informal process with no formal recommendations or written record,
- A formal process with formal recommendations and written responses to the recommendations, or
- Some variant in between.

One goal of the peer review or peer assist method is to have an independent set of eyes focusing on a company's operations with the sole purpose of helping that company improve. To ensure confidentiality, members of the team could be asked to sign a confidentiality agreement before serving. This method is based on the premise of promoting a "don't blame, let's improve" culture. The aviation industry is one in which the peer assist approach is employed.[1]

Key Performance Indicators

Key performance indicators (KPIs) are commonly used to evaluate a program's success or the success of a particular activity. KPIs work well when there are clear objective metrics that can be quantified, such as barrels of oil produced or number of lost-time incidents. A difficulty in using KPIs to assess the effectiveness of a SEMS program lies in determining the specific metrics that will be used to measure the effectiveness of the program. The process used by Petroleum Safety Authority (PSA) Norway, called *Risikonivå i norsk petroleumsvirksomhet*, is one approach that would be a useful starting point for BSEE KPIs. This approach is described more fully in the section on PSA Norway in Chapter 4.

Whistleblower Programs

A whistleblower program provides a means for an internal or external person (or organization) with knowledge that the SEMS program, or some of its components, is not being implemented correctly or is being

---

[1] See http://www.nasa.gov/offices/oce/appel/ask/issues/40/40i_peer_assist.html.

falsified to bring this information to the attention of the proper authority. In most cases such a program must protect the identity of the informant as well as guarantee that no repercussions, such as an employee's losing his or her job, will be forthcoming. Many industries use whistleblower programs, so there are many examples that can be used in conjunction with SEMS programs.

Periodic Lessee Reports

Operators or lessees may generate periodic reports describing the effectiveness of their SEMS program. Although perhaps open to questions about impartiality and accuracy, such reports do force the operator to take an active approach to SEMS implementation and monitoring. The contents of the report can range from an open format defined by the operator to a specific format and content required by the regulator.

Tabletop Exercises or Drills

Special drills or tests of an operator's SEMS program can be performed on a planned or surprise basis. Similar drills related to issues of life, safety, and environmental releases are already performed on offshore facilities. Because tabletop drills are not commonplace for SEMS, considerable planning by both the operator and the regulator would be needed to make the drill specific to testing the effectiveness of a SEMS program.

Monitoring Sensors

Mechanical sensors that monitor pressures, temperatures, flow rates, and related data can possibly be used in developing metrics that will determine the effectiveness of the SEMS program. The specific monitors, their relation to SEMS, and how such a system would work have yet to be determined. Some of these monitors may be in place already as part of normal production operations, while new monitoring devices specific to SEMS metrics may need to be developed. Ideally, these systems would be able to send information directly back to shore for real-time SEMS monitoring.

Calculation of Risk with SEMS in Place

A formal quantitative risk assessment (QRA) for a platform based on SEMS-specific data can be used to monitor the effectiveness of a SEMS program. The change in the QRA risk level when the SEMS program is modified or updated will show how effective the program is, although it is a computed theoretical effectiveness. One advantage of this method is that the owner can use the QRA risk level to determine the effectiveness of alternative SEMS-related modifications and upgrades to assist in determining the best approach from a SEMS perspective.

MEASURING TRENDS

The methods identified above directly assess the effectiveness of specific operator SEMS programs. However BSEE could aggregate the data across operators to monitor the trends and provide input to operators on specific improvements or areas of concern. Continuous improvement programs (CIPs), which are common in the offshore oil and gas industry, are one example of such an approach. In a CIP, employees typically submit suggestion slips or other forms of corporate feedback (sometimes anonymously) related to improvements to operations, including SEMS-type activities. Monitoring and reporting of these suggestions and how they change over time (e.g., an increasing or decreasing number of SEMS suggestions and the focus and types of suggestions) can be informative and lead to improvements in the industry's overall safety record. Another example is the industrywide collection and evaluation of SEMS-related data, such as data on safety and release incidents. Such data collections will improve the understanding of the effectiveness of SEMS across the industry as well as identify specific operators that have issues (or, conversely, that do not have issues) with their SEMS programs in comparison with their peers.

SUMMARY

Each of the methods described above could have a role in the assessment of both the progress being made in the implementation of SEMS and the effectiveness of SEMS. Evaluating SEMS is a continuous activity and therefore could include, at appropriate times and appropriate levels of the organization, a selection of the methods outlined above.

An audit is a periodic activity. Operating management, from first-line supervisors to top management, might find it useful to assess their progress toward improvement of safety and environmental conditions on an ongoing basis with a combination of SEMS monitoring sensors, KPIs, records of potential incidents of noncompliance, interviews, and other methods. Periodic assessment with drills, peer reviews, and lessee SEMS reports might have a broader scope than operational aspects and operating management. The methods that the committee recommends are presented in Chapter 6.

4

Existing Approaches for Assessing Safety Management Systems

> *It is impossible for a regulator to inspect quality into the petroleum industry. The industry itself must ensure quality.*
> —Magne Ognedal, Director General, Petroleum Safety Authority Norway

This chapter presents a description of the safety management programs of various U.S. and international regulatory agencies whose safety mandate is similar to that of the Bureau of Safety and Environmental Enforcement (BSEE). Each of these agencies has developed a program and established regulations to assure the compliance of the specific activities and cultures of the industries under its purview. In addition, the newly established Center for Offshore Safety (COS) is described, and its potential value to the U.S. offshore oil and gas industry is discussed.

The Committee on the Effectiveness of Safety and Environmental Management Systems for Outer Continental Shelf Oil and Gas Operations (the committee) received presentations from and conducted follow-up inquires with the Occupational Safety and Health Administration (OSHA) and Mine Safety and Health Administration (MSHA) of the U.S. Department of Labor, the U.S. Coast Guard (USCG), and the California State Lands Commission (CSLC), as well as with the United Kingdom (UK) Health and Safety Executive (HSE) and Petroleum Safety Authority (PSA) Norway. On the basis of the information gathered, the committee attempted to address the following questions for each agency, as applicable:

- What has been done to ensure there is a safety management system (SMS) in place?
- How does the regulatory authority know that the SMS is working?
- How does the regulatory agency enforce it?

- Now that the agency has had some experience with a safety management program, what does it believe is effective in the program? What would the agency change in the program if it could?

U.S. REGULATORY AGENCIES

U.S. Coast Guard

The USCG policy for enforcing the International Safety Management (ISM) Code is divided into two major areas. The first area of responsibility is for U.S. flag vessels mandated to comply with the ISM Code. USCG is the flag administration agency for the implementation and enforcement of the ISM Code on U.S. flag vessels and administers this responsibility through a delegation of recognized and authorized organizations according to 46 CFR 8, "Vessel Inspection Alternatives." The second area of responsibility is verification of compliance with the ISM Code on foreign-flag vessels entering U.S. ports. Detailed guidelines for enforcement of the ISM Code on foreign-flag vessels subject to the U.S. Port State Control program are contained in Navigation and Vessel Inspection Circular (NVIC) 4-98. This NVIC contains all of the applicable International Maritime Organization guideline documents for the ISM Code.

Compliance with the ISM Code is unique because the code is integral to nearly every other aspect of overall regulatory compliance. A basic tenant of any SMS is that the system must be in constant compliance with requirements for safety and environmental protection. Because of this, USCG marine inspectors will, in the course of routine material and human element inspections, provide a means of measuring compliance with the ISM Code. Confirmation of compliance can take several forms, the most basic of which is simply to verify that the vessel has a valid ISM Code Safety Management Certificate and a copy of the company's Document of Compliance Certificate. The next, and more complex, level is to identify links between any deficiencies noted during the course of routine inspections and the vessel's SMS. The latter task requires that marine inspectors have a working knowledge of the elements of the ISM Code as well as knowledge of the duties and training of shipboard personnel. To assist marine inspectors in making these judgments, a training course has been established at the Marine Inspection and Investigation School at the USCG Training Center in Yorktown, Virginia. All USCG marine inspectors and vessel-boarding officers are required to read and become familiar with the ISM Code and NVIC 04-98.

USCG oversight of ISM Code auditing or ISM Code certification processes for the SMS of a U.S. company or vessel is coordinated through the authorized organization. Any examination of a vessel for any purpose is also an opportunity to judge the effectiveness of the SMS. Although ISM oversight is not the primary purpose of the visit, inspectors are alert to whether the deficiencies they find while performing other inspections should have been managed with the SMS. Oversight may also occasionally arise from investigations into vessel casualties, reports by vessel crew members, or at the direction of the USCG commandant.

Any time an authorized organization's surveyor notes significant material deficiencies, serious lack of maintenance of a vessel or its equipment, or failure of the crew to follow safety procedures, the potential or actual failure of the SMS procedures is analyzed. This analysis may include instances of a lack of routine maintenance of critical systems or of equipment or material failures that have not been submitted as a corrective action request and that indicate a clear failure of the crew to follow maintenance or safety procedures. Information to make this type of determination may be collected by

- Observing or interviewing the crew members responsible for the area of the SMS where the deficient item was noted. Crew members should be knowledgeable about the responsibilities required of them by SMS procedures.
- Verifying that SMS procedures are being carried out with regard to the area of deficiency.
- Asking the master or responsible crew member to give an account of what corrective action has been initiated under the SMS and to cite evidence of this action. Failure to submit corrective action reports is noted and, depending on the severity and number of instances, is reported to the organization that issued the Safety Management Certificate. When these failures are found, the representative of the authorized organization acting on behalf of the United States must provide a report, orally or in writing, to the cognizant local officer in charge, marine inspection. These reports are required to be submitted as soon as possible; in addition, oral or written reports (the latter of which can be delivered via e-mail) are supposed to be made within 48 hours.

If it appears that any portion of the SMS is not being followed, USCG personnel may issue a vessel deficiency citation (Form CG-835, Notice of Merchant Marine Inspection Requirements) to the vessel's master requesting verification of compliance from the authorized organization that issued the vessel's Safety Management Certificate. If the nonconformity is linked to shoreside operations, then compliance from the authorized organization that issued the company's Document of Compliance Certificate is also required. It is the vessel master's responsibility to notify the organization that issued the Safety Management Certificate or Document of Compliance Certificate. Depending on the severity of the deficiency, USCG may allow a reasonable period of time to satisfy the requirements of the CG-835. In cases in which the deficient item would restrict the vessel from sailing, the time allowed by the CG-835 for verification of the SMS should be proportionally short.

U.S. Occupational Safety and Health Administration

Process safety management (PSM) is an OSHA regulation intended to prevent or minimize the consequences of a catastrophic release of hazardous materials from specific onshore processing operations, notably chemical and hydrocarbon facilities. PSM is similar to Safety and Environmental Management Systems (SEMS) in that it involves comprehensive procedures and management practices to ensure safe operations that protect workers and, by extension, minimize environmental consequences. The PSM rule is contained in 29 CFR 1910.119, "Process safety management of highly hazardous chemicals."

PSM was initiated in 1992 after several large-scale chemical incidents, including the explosion in Flixborough, England, in 1974; the toxic release in Bhopal, India, in 1984; the toxic release at the Union Carbide facility in Institute, West Virginia, in 1985; and others. Investigations and studies of these events indicated that a performance standard was needed that would provide a comprehensive management program—a holistic approach that would integrate technologies, procedures, and management practices. The details of such a program are contained in 29 CFR 1919.119.

PSM covers 225 different industry subsectors with an estimated 10,000 to 15,000 processes. The PSM regulation (29 CFR 1910.119) was first published in February 1992. Covered facilities were required to comply with the standard by May 26, 1992. The standard provided a period of approximately 5 years for employers to conduct their initial process hazard analyses (PHAs), with 25 percent of the PHAs to be conducted in each year, starting in 1994, and all PHAs to be completed by May 1997. The PHA element of PSM must be updated and revalidated at least every 5 years [29 CFR 1919.119 (c)(6)], and audits to ensure compliance with all provisions of PSM must be conducted at least every 3 years [29 CFR 1919.119 (o)].

In contrast, the SEMS regulation was published in October 2010, with full implementation required by November 15, 2011 [30 CFR 250.1900 (a)]. Although November 2011 was the deadline for implementation of a SEMS plan, operators were not required to submit a written plan. Instead, they have been subject to audit at any time thereafter and must be able to demonstrate they have a SEMS plan in place if there is an incident.

Several American Petroleum Institute (API) publications that address PSM with regard to oil and gas operations are available, including API Recommended Practice (RP) 750, *Management of Process Hazards* (API 1990); API RP 752, *Management of Hazards Associated with Location of Process Plant Buildings* (API 2003); API RP 754, *Process Safety Performance Indicators for the Refining and Petrochemical Industries* (API 2010b); and API RP 755, *Fatigue Risk Management Systems for Personnel in the Refining and Petrochemical Industries* (API 2010a). The latter two incorporate recommendations from the study of the 2005 Texas City explosion (ABSG Consulting Inc. 2006; BP U.S. Refineries Independent Safety Review Panel 2007). PSM also references several publications related to chemical plants and other types of industrial facilities that handle hazardous materials. Because PSM is a performance management standard, it requires employers to identify the codes and standards they use with respect to equipment and to document that they have complied with recognized and generally accepted good engineering practices for the design, inspection, and testing of their equipment.

For offshore oil and gas operations, SEMS likewise references API RP 75, *Recommended Practice for Development of a Safety and Environmental Management Program for Offshore Operations and Facilities* (API 2004). BSEE used the Safety and Environmental Management Program (SEMP) as the underlying philosophy for SEMS. BSEE informed the committee that they conferred with the OSHA PSM group while developing SEMS in order to incorporate lessons learned and other findings from OSHA's approximately 20 years of experience with PSM.

The initial PSM rule had 12 elements; two more elements (employee participation and trade secret protection) were added later. The PSM elements are similar to the SEMS elements (see Table 4-1). There are

TABLE 4-1 Comparison of SEMS Elements with OSHA PSM Elements

SEMS Element (CFR reference)	Similar OSHA PSM Element (PSM element number)	General Description
1. General (30 CFR 250.1909)		Implementation, planning, and management approval of program
2. Safety and environmental information (30 CFR 250.190)	Process safety information (2)	Compilation of written process safety and environmental information, including hazard information
3. Hazards analysis (30 CFR 250.1911)	Process hazards analysis (3)	Conduct of PHA for each covered process
4. Management of change (30 CFR 250.1912)	Management of change (10)	Establishment and implementation of written procedures to manage change
5. Operating procedures (30 CFR 250.1913)	Operating procedures (4)	Development of written operating procedures that provide clear instructions for safely conducting activities
6. Safe work practices (30 CFR 250.1914)	Hot work (9) Line breaking (4) Lockout–tagout (4) Confined space entry (4)	Development and implementation of practices for hazardous operations
7. Training (30 CFR 250.1915)	Training (5) Contractors (6)	Conduct of training of employees and contractors alike; training must emphasize safety and environmental hazards
8. Mechanical integrity (30 CFR 250.1916)	Mechanical integrity (8)	Development of written procedures for the ongoing integrity of process equipment
9. Pre-start-up review (30 CFR 250.1917)	Pre-start-up safety review (7)	Confirmation that the construction and equipment of a process are in accordance with design specifications
10. Emergency response and control (30 CFR 250.1918)	Emergency planning and response (12)	Development and implementation of an emergency action plan

(continued)

TABLE 4-1 *(continued)* Comparison of SEMS Elements with OSHA PSM Elements

SEMS Element (CFR reference)	Similar OSHA PSM Element (PSM element number)	General Description
11. Investigation of incidents (30 CFR 250.1919)	Incident investigations (11)	Investigation of each incident that resulted in, or could reasonably have resulted in, an incident
12. Auditing (30 CFR 250.1920)	Compliance audits (13)	Evaluation of the program of compliance
13. Record keeping (30 CFR 250.1928)		Maintenance of documentation that describes the elements of the program
14. Stop work authority[a] (30 CFR 250.1930)		Stipulation that any and all personnel (employees or contractors) can stop unsafe or hazardous work
15. Ultimate work authority[a] (30 CFR 250.1931)		Identification of the person with ultimate authority for the facility
16. Employee participation[a] (30 CFR 250.1932)	Employee participation (1)	Development of a written plan of action regarding the implementation of employee participation
17. Guidelines for reporting unsafe work conditions[a] (30 CFR 250.1933)		Provision of procedures to report unsafe work conditions
18. None	Trade secrets (14)	Information required by the PSM standard is to be made available as needed (confidentially if necessary)

[a]Additional element issued under SEMS II in September 2011 (BOEMRE 2011a).

13 original SEMS elements, and several more were proposed in the notice of proposed rulemaking published in September 2011 (SEMS II) (BOEMRE 2011a).

Early PSM compliance used a program quality verification scheme in which compliance safety and health officers audited an operation for PSM compliance and OSHA issued citations for noncompliance. Program quality verification was resource intensive, although relatively few citations were issued, and was too broadly focused. It did not focus the compliance safety and health officers on specific issues for the many types of facilities and processes covered by PSM (in contrast, SEMS is generally limited to offshore oil, gas, and sulfur operations). Program quality verification was subsequently replaced with the current National Emphasis

Program for PSM enforcement. This system uses a list-based approach for determining compliance via a publicly available "static list" of compliance items to be inspected and a "dynamic list" that is not publicly available and is ever changing. Because the National Emphasis Program is able to conduct more inspections with the same number of resources, there is more incentive for better compliance with the standard. The National Emphasis Program has uncovered many more significant findings than the previous program quality verification approach. This is partly because of the large-scale refining and chemical facilities and operations to which PSM applies, as it is easier to identify deficiencies when there is a focus on specific items to evaluate. OSHA has also increased PSM training for compliance safety and health officers in order to provide a more effective workforce.

Discussion between the committee and OSHA identified the following actions that could be taken to improve PSM:[1]

- Revise the wording of the PSM regulations to make them more defensible against legal arguments that try to work around the phrasing in the Code of Federal Regulations. OSHA believes that the PSM requirements are not fundamentally flawed; rather, some modifications in the wording of the requirements would improve the ability to enforce them.
- Look at specific performance requirements to determine whether they can be made more prescriptive without burdening the employer.
- Use dedicated staff with experience and background in the industry that is being inspected.
- Emphasize the need for comprehensive training for enforcement staff and managers.

Additional information about PSM can be found on the OSHA PSM website.[2] This website provides the PSM regulations and references for equipment design and in-service practices (including inspection, testing, preventative and predictive maintenance, repair, alteration, rerating, and

---

[1] M. Marshall. OSHA's PSM Regulation. Presentation to the committee, Washington, D.C., August 31, 2011.
[2] See http://www.osha.gov/SLTC/processsafetymanagement/index.html.

fitness-for-service evaluations). The website also covers other important aspects of PSM, including PHA, human factors, facility siting (blast), fire protection, mechanical integrity, procedures, management of change, and other issues. An extensive list of references provides access to other PSM-related information.

Mine Safety and Health Administration

MSHA was created in the U.S. Department of Labor in 1977 with the passage of the Federal Mine Health and Safety Act of 1977 (the 1977 Mine Act) and has responsibility for enforcing safety and health rules in all mines and mineral-processing operations in the United States, regardless of their size, the number of employees, or the method of extraction. The Mine Act provides that MSHA inspectors shall inspect each surface mine at least two times a year and each underground mine at least four times each year (seasonal or intermittent operations are inspected less frequently) to determine whether there is compliance with health and safety standards or with any citation, order, or decision issued under the Mine Act; to determine whether an imminent danger exists; and to ensure that actual practices comply with approved plans and programs.[3] Some of MSHA's other important mandatory activities are

- Reviewing for approval mine operators' mining plans and education and training programs,
- Developing improved mandatory safety and health standards,
- Investigating petitions for modification of mandatory safety standards,
- Investigating mine accidents and hazardous condition complaints, and
- Assessing and collecting civil monetary penalties for violations of mine safety and health standards.

To fulfill its mandate, MSHA currently has approximately 800 coal inspectors and 345 metal and nonmetal inspectors; the agency also has more than 200 full-time exempt employees in support of its technical support function. In FY 2011, MSHA had 2,200 full-time exempt employees and a budget of approximately $380 million.

---

[3] See http://www.msha.gov/REGS/ACT/ACTTC.HTM.

MSHA pursues several other activities that support the carrying out of the mandates of the 1977 Mine Act. Some of the important activities are associated with the education and training of mine inspectors, mine officials, and miners; the testing, approval, and certification of certain mining products for use in mines; and the provision of technical assistance to the states and small mine operators. These are accomplished through specific mechanisms such as MSHA's

- National Mine Health and Safety Academy,
- Approval and Certification Center,
- Pittsburgh Safety and Health Technology Center, and
- Directorate of Educational Policy and Development.

The partial list of MSHA's mandatory and support activities given above points to a number of opportunities for auditing the status of health and safety in mines. In addition to mandatory inspections, strategic impact inspections at mines that may need greater attention are also conducted. Such mines could be characterized as having a high risk of explosion; a poor history of compliance; or a high incidence of injuries, illnesses, fatalities, violations, or complaints.

Auditing in MSHA's Approval and Testing Center encompasses a large number of verification, validation, and approval processes. Of particular importance is the postapproval audit by the agency. Review and approval of mine operators' mining plans, training programs, and certification of trainers and mine officials provide the basis for verification and validation during inspections and audits.

MSHA has one of the most comprehensive computerized databases of mining operations and mine health and safety statistics in the world. The National Institute for Occupational Safety and Health (NIOSH) has converted this database into two popular formats—dBase IV and SPSS (which includes labels and coding information)—so that NIOSH and other interested parties (including consultants, universities, and the National Safety Council) can access and analyze MSHA data in the course of researching and advancing health and safety experiences in the mining industry.

MSHA's work with NIOSH, industry, and states to develop health and safety programs is extensive. States and trainers use MSHA's State

Grants Program for miner training programs and MSHA's training resource materials to conduct health and safety training.

MSHA also has an alliance with the National Stone, Sand, and Gravel Association (NSSGA)—a sand, stone, and gravel operators group—to help in the development and implementation of health and safety programs to create a culture that will prevent accidents and injuries in these mines. The NSSGA–MSHA Alliance has defined, described, and developed examples of 11 fundamental principles of a safety program that covers elements such as management commitment, training and development, employee involvement and participation, incident investigation, and accountability.

Recently, MSHA has undertaken a new rule-making process to implement new regulations for safety and health management programs in the mines. MSHA believes that operators with effective safety and health management programs will identify and correct hazards more quickly and successfully, which will reduce the number of accidents, injuries, and illnesses. In October 2010, MSHA held three public information-gathering meetings. Information received from these meetings indicates that companies with a safety and health management program have better health and safety records.

MSHA is still gathering data to determine what actions the agency might take, including implementation of specific regulations governing safety and health management programs. For example, to gather information on existing model programs for best practices for safety and health management programs, MSHA held a public meeting on November 10, 2011, in Birmingham, Alabama, in conjunction with the Sixth Annual Southeastern Mining Safety and Health Conference. Proposed rules are expected sometime in 2012 and may be similar to the Injury and Illness Prevention Programs being proposed by OSHA.[4]

In summary, MSHA has an extensive program encompassing compliance inspections, impact inspections, equipment testing and approval, review and approval of mine plans, compliance assistance, education and training programs, trainer and official certification, and technical services, all of which may provide some insights to BSEE.

---

[4] L. Zeiler, United States Department of Labor, Mine Safety and Health Administration. Presentation to the committee, Washington, D.C., August 31, 2011.

California State Lands Commission

CSLC requires operators to comply with API RP 75 (SEMP) and conducts a program called the Safety Assessment of Management Systems (SAMS) that conducts an external independent safety audit of California's oil and gas facilities every 5 years. SAMS is based on a joint industry project performed in the 1990s by Paragon Engineering Services with assistance from the University of California, Berkeley, and sponsored by the Minerals Management Service (MMS), CSLC, HSE, the National Energy Board of Canada, the American Bureau of Shipping (ABS), Chevron, and Texaco. CSLC has been using SAMS to audit API RP 75 SEMP performance for more than 15 years. This technique was originally developed by the joint industry project and modified slightly with experience; CSLC has used SAMS to review some installations three times over the years.

CSLC also conducts a physical condition, design, and safety compliance audit on the same 5-year interval as SAMS. This audit complements the inspection program and provides for strong familiarity with the facility before the SAMS audit is conducted. CSLC believes that a hardware-oriented inspection or audit program does not address the SMS, human factors, or safety culture, and the commission saw the need for these new types of audits more than a decade ago. CSLC has observed steady improvement in safety management performance and culture using the SAMS process and attributes these improvements to working with operators to increase their compliance rather than punishing them with fines and shut-ins for areas that may need improvement. In affording operating companies the latitude to develop their programs, CSLC has observed that several operators have made great strides in using behavioral safety observations to identify areas for improvement and in fostering improved safety culture among their employees.

CSLC staff try both to work closely with operators to improve safety culture and to avoid the perception of being adversarial regulators. They ride the company crew boat with company personnel; attend company-required facility safety orientations and morning safety meetings; and observe the actual use of work permits, prejob safety reviews, lockout–tagout procedures, and company operations. They also discuss the SAMS programs with the people that implement and use them and observe improvements that occur as a result. Firsthand knowledge of general maintenance conditions, work processes, maintenance backlog, and the number of sensors in bypass are additional qualitative performance measures and indicators that are employed.

In essence, the emphasis is on promoting a culture of safety, from senior management down to the rig floor workers, rather than on safety compliance. As a safer culture develops, CSLC inspectors have noted that operator staff who have participated in CSLC's behavioral safety observation programs appear to take more pride in their work and are willing to describe how their programs have evolved and improved.[5] In CSLC's view, these are some of the elements that have helped them be successful where other regulatory, corporate, and even third-party paperwork audits have failed.[6] Verified documentation does not equate to a true implementation of a positive safety culture, but working closely with operator staff appears to do so.[7]

INTERNATIONAL REGULATORY ORGANIZATIONS

UK Health and Safety Executive

The current UK offshore regulatory goal-setting regime was born out of a public inquiry into the *Piper Alpha* explosion in 1988. The goal-setting legislation replaced older prescriptive legislation, and HSE replaced the UK Department of Energy as regulator. HSE set up the Offshore Division, which has two types of inspectors: regulatory management inspectors and specialist or topic inspectors. Topic inspectors specialize in areas such as process safety; mechanical, electrical, and marine issues; and occupational health. They provide in-depth assessments of safety cases and input into offshore inspections and investigations. Regulatory management inspectors manage the assessment of safety cases and lead offshore inspections and investigations, with the participation of topic inspectors.

---

[5] M. Steinhilber, California State Lands Commission. Presentation to a subgroup of the committee, Long Beach, California, March 24, 2010.
[6] M. Steinhilber, California State Lands Commission. Presentation to a subgroup of the committee, Long Beach, California, March 24, 2010.
[7] M. Steinhilber, California State Lands Commission. Presentation to a subgroup of the committee, Long Beach, California, March 24, 2010.

As a regulator, HSE engages with industry at all levels. HSE influences a duty holder's (operator's) senior management by meeting with them in other forums; participates in industry committees, workgroups, joint industry research, and conferences; and, finally, conducts regular offshore inspections and investigations. During offshore visits, inspectors engage with offshore management and the workforce through formal and informal interviews and discussions to seek evidence of compliance. Inspectors usually spend 3 days offshore. They travel to an installation via the regular scheduled helicopter flights that serve it and do not pay for meals or overnight stays while offshore. The duty holder is invoiced for the inspectors' time.

To ensure there is an SMS in place, the UK offshore regulatory system requires companies that operate production installations and those that own nonproduction installations (e.g., drilling rigs)—both referred to as "duty holders"—to submit a safety case to HSE for assessment and acceptance prior to operation of an installation. The Safety Case Regulations 2007 require several specific items to demonstrate that the management system is adequate:

- Compliance with the relevant statutory provisions with respect to matters within the management system's control;
- Satisfactory management arrangements with contractors and subcontractors;
- Established adequate arrangements for audits and for making reports thereof;
- Identification of all hazards with the potential to cause a major accident; and
- Evaluation of all major risks and implementation, or plans for implementation, of measures to control those risks to ensure compliance with the relevant statutory provisions.

HSE assesses the evidence provided in the safety case and, if the evidence is deemed sufficient, accepts the case. The duty holder is then allowed to operate the installation. If the safety case is not accepted, operation of the installation would be illegal. During the assessment process, HSE often identifies weaknesses in a case and discusses with the duty holder the required additional information.

HSE undertakes planned inspections covering a range of topics and issues within the safety case and checks compliance with all relevant statutory provisions to determine whether the SMS is working. These inspections involve testing the effectiveness of the duty holder's SMS as applied on the offshore installation. The system is also tested when HSE investigates accidents and incidents. Inspections and investigations involve checking the duty holder's policies and procedures; examining records and other documents that are a product of the system (e.g., maintenance records); and speaking to onshore and offshore duty holder managers and members of the offshore workforce to seek evidence of their understanding of the management system and its application to specific work activities or operations and practices. On occasion, HSE inspectors formally record interviews as formal statements when HSE is undertaking an investigation or anticipating a formal enforcement action. Inspectors observe work activities and, from time to time, seek demonstration of the effectiveness of particular equipment. For example, an inspector might require the deluge system in a module to be operated to check for blocked nozzles and water spray coverage and to ensure that it meets performance standards.

HSE has a public enforcement policy[8] that is supported by a guide known as the Enforcement Management Model.[9] HSE has a range of tools for enforcement and applies these in a proportionate and targeted way; however, evidence of breaches of legislation is required before any enforcement steps are taken, and then inspector judgment must be used in applying the enforcement policy. The tools and approaches available include

- Serving a duty holder or employer an improvement or prohibition notice,
- Directing a duty holder to revise a safety case, and
- Prosecuting the duty holder or other employer.

An improvement notice is served when an inspector believes that a breach of legislation has occurred and that it would be appropriate to serve such a notice. The notice explains the breach and provides the date

---

[8] See http://www.hse.gov.uk/pubns/hse41.pdf.
[9] See http://www.hse.gov.uk/enforce/emm.pdf.

by which the duty holder must comply. A schedule describing the actions the duty holder should take to achieve compliance is often attached to the notice. Duty holders, however, are not required to follow the proposed actions; they are entitled to take any other effectively equal measures to achieve compliance. HSE then visits the offshore facility to determine whether the duty holder has complied with the notice. If the original date for compliance becomes unrealistic for genuine reasons, the duty holder may seek an extension.

A prohibition notice is served when the inspector believes there is evidence that an activity or operation will lead to serious injury. The notice describes the operation or activity and the circumstances that give rise to the risk of serious personal injury. When the notice is served, the activity or operation that gives rise to the identified risk must be changed. A schedule similar to the improvement notice schedule may be attached.

Improvement and prohibition notices are both legal documents; thus, the duty holder can appeal them to question the inspector's reasoning and evidence. An appeal is heard in an employment tribunal. When an improvement notice is appealed, the duty holder does not have to take any steps to comply with it until the tribunal renders a decision. If the appeal fails, the duty holder must comply with the notice.

When a prohibition notice is appealed, the prohibited activity must stop and may only start up again if the tribunal rules in the duty holder's favor. Tribunals, which can take months to arrange, can last from several days to many weeks. Appeals against notices do happen, but they are infrequent. Failure of a duty holder to comply with a notice is a prosecutable offense. The Offshore Division serves about 35 to 50 improvement notices and two to three prohibition notices each year.

In consideration of the *Deepwater Horizon* accident, the UK government is currently undertaking a review of the existing health, safety, and environmental regime for the UK Continental Shelf. The report of the findings of this review is expected to be released later this year.

PSA Norway

PSA Norway has regulatory responsibility for the technical and operational safety of petroleum activities, including emergency preparedness and the working environment (Norwegian Petroleum Directorate 2011,

p. 18). PSA reports through the Ministry of Labor, while a sister agency, the Norwegian Petroleum Directorate, has responsibility (e.g., leasing, collection of royalties) for developing Norway's petroleum resource and reports through a separate Ministry of Petroleum and Energy.

PSA was established in 2002 when the government split it off from the Norwegian Petroleum Directorate (which at that time was responsible for both safety and petroleum resource development) to form two separate agencies, each reporting to different ministries. The new PSA assumed responsibility for safety on all offshore petroleum facilities and those onshore facilities associated with offshore petroleum production. PSA takes a holistic approach to the meaning of safe operations and extends the concept of safety to include protection of human life, health, and welfare; the natural environment; financial investment; and operational regularity.

PSA currently regulates more than 40 operating companies with 70 offshore production facilities, two onshore facilities, 14,000 km of pipelines, and about 30 floating and 12 platform drilling rigs on the Norwegian Continental Shelf (Norwegian Petroleum Directorate 2011, p. 25; PSA Norway 2011d). About 25,000 people work offshore in Norway. The great majority of offshore operations in Norway are at water depths of less than 350 m, although one field (Orme Lange) operates at a water depth of about 1,100 m.

To regulate and audit these operations, PSA employs about 166 staff, of which about 110 are professionals. Sixty of the professional staff are qualified audit team leaders [qualifications were initially based on ISO 9000 (ISO 2005), but it was not specific enough, and requirements and training have since been improved].[10] PSA personnel are compensated at a level about two-thirds that of personnel with comparable Norwegian petroleum industry responsibilities.[11]

The Norwegian petroleum sector's initial approach to regulation was based on the assumption that the oil companies were not capable of performing safely without strict regulatory policing (PSA Norway 2010). The initial approach was to establish prescriptive laws and regulations

---

[10] M. Ognedal, PSA Norway. Presentation to the committee, Houston, Texas, October 19, 2011.
[11] M. Ognedal, PSA Norway. Presentation to the committee, Houston, Texas, October 19, 2011.

that set specific requirements for structures, technical equipment, and operations in order to prevent accidents and hazards. Quite comprehensive inspections of facilities and activities were conducted with detailed regulatory punch lists, with the goal of ensuring compliance. At one point, regulations required quantitative calculation of the maximum acceptable risk that an accident would occur and specified that it should not be greater than a statistical probability of 1 in 10,000; however, experience with this approach, including several blowouts and several high-profile tragedies—most notably the loss of 123 lives on the *Alexander L. Kielland*—was not as desired.[12]

With quantitative risk calculation, it was found that discussions on the risk requirements for approving new developments on the Norwegian Continental Shelf quickly became pure number-crunching exercises. That, in turn, meant it was easy for statisticians to document that the various risks in such projects were within the acceptable limits (PSA Norway 2010).

Furthermore, prescriptive, detailed compliance inspection was found to have encouraged a passive attitude among companies. They waited for the regulator to inspect, identify errors or deficiencies, and explain how these were to be corrected. As a result, the authorities became, in some sense, a guarantor that safety in the industry was adequate, and they thereby assumed a responsibility that should have rested with the companies (PSA Norway 2010).

These limitations led Norway, over time, to move from prescriptive to performance-based regulation, which involves specification of the performance or function to be attained or maintained by the industry (PSA Norway 2010). The regulatory role involves defining the safety standards that companies must meet and checking that they have management systems that ensure such compliance. For their part, the companies are given a relatively high degree of freedom in selecting good solutions that fulfill the official requirements.

The term "inspection" was replaced with the preferred term "supervision," and "approvals" was replaced with "consents." PSA believed the

---

[12] M. Ognedal, PSA Norway. Presentation to the committee, Houston, Texas, October 19, 2011.

changes in terminology were more significant than might be thought at first sight. The supervision concept, for instance, was not confined to mere monitoring. It covered all activities that provided the necessary basis for determining whether the companies had accepted responsibility for complying with the regulations in every phase (PSA Norway 2010). By way of simple analogy, under quality management, the supervisor of a factory floor does not just inspect the factory's work product and fire employees who do not perform. Rather, the supervisor works with employees to ensure performance continuously improves.

The change in philosophy created a climate in which PSA worked with the industry to improve safety instead of acting in the role of a compliance inspector and guarantor of the acceptability of company activities. In the context of PSA, supervision is directed at the operator's administrative management system, which the companies actually use to ensure acceptable operation.

PSA works with individual operators with the intention of helping to make them more successful, but also works with the industry by chairing an industry board that consists of representatives from employers' representatives (operators, manufacturers, and shipping associations), employees (represented by five unions) and regulators to define the regulations and issue nonbinding recommendations and guidelines. PSA works very closely with employers and employees, but PSA ultimately makes the decisions. These recommendations and guidelines make frequent reference to international industrial standards for equipment, structures, and procedures.

To confirm that there is an SMS in place, PSA conducts audits of companies to ensure acceptable operation. These audits are conducted by personnel with the special expertise and experience necessary. During the audit, the operator must demonstrate both a commitment to and an expertise in complying with the frame conditions that govern its operations. According to PSA, the requirements of a performance-based system audit place a great demand on industry, employees, and the regulator in terms of expertise, management, and flexibility.[13]

---

[13] M. Ognedal, PSA Norway. Presentation to the committee, Houston, Texas, October 19, 2011.

A typical audit is conducted by a team of at least two but up to five or six people. From planning through execution and reporting, the audit takes 2 to 5 weeks. Notice is usually given about a month in advance, and separate meetings are held with union safety delegates to ensure that employee views are heard.

The audit team and plan vary according to the type of operation being audited. Detailed audit guidelines are designed for each audit, and each audit team is led by a certified audit team leader. Audit team membership is driven by the competencies needed to perform the task. For example, if maintenance management is of particular interest, the team will include a maintenance specialist.

Scheduling of audits is not determined solely by frequency, but by using a risk-based approach. Operations and particular operators are chosen for audits based on risk. In addition to the risk-based audits, an audit of each installation is conducted at least once every 3 years.

For the purpose of integrity, there are never fewer than two auditors present. Norway, however, is a comparatively small country (4.8 million residents; in comparison, Louisiana has 4.5 million residents) with a relatively large petroleum industry (the industry accounts for more than 20 percent of the country's gross domestic product). Therefore, no attempt is made to limit audit team membership on the basis of prior or current involvement of a family member or friend in the organization being audited. Norway has not felt a need to institute detailed conflict of interest regulations beyond direct financial involvement with the operator.[14]

Either the operator or an independent designee conducts inspections (both independent and internal) as a normal part of the management system. PSA may or may not request the results of these inspections as part of its management system audit. In addition to audits, PSA conducts incident investigations with special focused teams as necessary. The results of these investigations are used to improve operations in the investigated operator's organization and to inform and improve operations in organizations with similar kinds of operations.

PSA uses several formal instruments other than audits and inspections to assess how well an operator's SMS is working (PSA Norway 2011d).

---

[14] M. Ognedal, PSA Norway. Presentation to the committee, Houston, Texas, October 19, 2011.

Although structured, a few of these instruments are, to some extent, quantitative:

- *Dialogue.* The bulk of practical supervision consists of dialogue between PSA and the industry to assess trends and request plans, analyses, documentation, and information. Meetings between PSA and the relevant company involve both appropriate managers and employees. Generalized results that do not identify specific operators or installations are summarized in reports that are then posted to PSA's website.[15]
- *Notification of undesirable incidents.* Companies are required to notify PSA about undesirable incidents. The regulations clearly define what must be reported and require the use of a dedicated form. Approximately 800 to 900 notifications are received every year. The number of undesirable incidents and the character of these incidents also help PSA assess an operator's management system. An abnormally low number of incidents may indicate a problem with an operator's reporting system and, therefore, with the operator's entire SMS. An abnormally high number of incidents may indicate a safety problem.
- *Hotline.* PSA has a hotline staffed around the clock for reporting emergencies. Such reports are first received and registered by the duty officer, who also makes the initial assessment of the seriousness of the incident and the possible immediate actions required. If necessary, the duty officer activates PSA's emergency response center to monitor a serious ongoing incident.
- *Tailored follow-up.* Each undesirable incident is allocated to a case officer who checks it, categorizes its seriousness (which may differ from the operator's assessment), and selects a tailored follow-up for the operator or company. In the case of very serious events, PSA may decide to launch an investigation or conduct another type of close follow-up. The response to less serious incidents is tailored to their nature.
- *Risikonivå i norsk petroleumsvirksomhet (RNNP).* The RNNP process was initiated in 1999–2000 to develop and apply a tool for measuring trends in risk level in Norwegian petroleum activity. This process monitors risk trends with the aid of various methods, such as incident indicators, barrier data, interviews with key informants, working

---

[15] See http://www.ptil.no/main-page/category9.html?lang=en_US.

seminars, and fieldwork. A major questionnaire-based survey is also conducted every 2 years. This work has acquired an important position in Norway's oil and gas industry because it contributes to a shared understanding of risk developments on the part of everyone involved. Results from these studies are presented in annual reports, which also provide the basis for taking action to combat a negative trend. Published around April 20 each year, the annual reports provide a realistic picture of developments in the risk level from year to year. The RNNP process only indirectly helps assess a particular operator's management system, but does provide trend information that helps PSA take action.

- *Whistleblowers.* Whistleblowers help to shape and complete the picture of the safety position. PSA receives information from employees in the industry about poor safety or conditions open to criticism in their workplace. Such input is closely followed up in accordance with established and legally required routines. The anonymity of whistleblowers is protected. Whistleblowers help PSA understand how a particular company's management system is working by identifying possible issues that are not found by other means.
- *Daily Drilling Report System (DDRS).* Since 1984, companies have been required to provide information via the DDRS database on all drilling on the Norwegian Continental Shelf. PSA can extract from the DDRS the essential facts about each current operation and thereby assure itself, if necessary, that undesirable well incidents have actually been reported.

PSA applies the necessary measures to ensure that activities are conducted in accordance with regulatory requirements and through formal instruments.[16] These measures include the following:

- Observations with comments, which are discussed with the operator;
- Improvement possibilities, which drive discussion and are reported to the operator (the operator is required to inform PSA of the changes made as a result);
- Issuance of a "not in compliance" notice, with a requirement to fix the problem in less than 3 weeks;

---

[16] M. Ognedal, PSA Norway. Presentation to the committee, Houston, Texas, October 19, 2011.

- Police investigation for willful violation (which has happened once);
- Recommendation to the ministry to remove the operator's license; and
- Recommendation that the operator be banned from future drilling blocks in Norway.

It is PSA's responsibility to define the terms for responsible operation of the petroleum industry and to check that companies are working on prevention and on continuous improvement of safety levels. Because PSA views criminal law as the province of the police (PSA Norway 2010), almost all PSA enforcement actions are in the form of observations or improvement possibilities.

PSA believes that the following aspects of its program (presented in no particular order) are critical to its program's effectiveness:

- Doing nothing to take responsibility away from the industry. The PSA model is based on the conviction that the government cannot inspect quality into the industry. The industry itself must ensure that quality is achieved and maintained (PSA Norway 2010).[17]
- Dialogue on problems. PSA believes in working with operators and the industry to make them more successful. The internal control system can only work as intended if it is operated in close collaboration and consultation with safety delegates, employees, and the regulator (PSA Norway 2010).[18]
- A focus on functional requirements and system orientations rather than on checking compliance.[19]
- A "fit-for-purpose" approach to constituting audit teams. Teams must consist of sufficient personnel with the expertise and experience necessary for a specific audit.[20]
- The RNNP approach, which provides flexibility and focus to supervision (PSA Norway 2010).
- Allowing operators, to a great extent, to choose for themselves the solutions they will adopt to meet official requirements (PSA Norway 2011c).

---

[17] M. Ognedal, PSA Norway. Presentation to the committee, Houston, Texas, October 19, 2011.
[18] M. Ognedal, PSA Norway. Presentation to the committee, Houston, Texas, October 19, 2011.
[19] M. Ognedal, PSA Norway. Presentation to the committee, Houston, Texas, October 19, 2011.
[20] M. Ognedal, PSA Norway. Presentation to the committee, Houston, Texas, October 19, 2011.

- Involvement of PSA specialists in both monitoring and participating in the development and revision of industrial standards to help make sure that these are constantly relevant and reflect best practices (PSA Norway 2011c).
- Recognition that the work involved in a performance management system can easily be underestimated, and that it is therefore important to emphasize that this form of regulation demands much more of the industry, employees, and government than detailed regulations (PSA Norway 2011c).

PSA uses the RNNP process and the past performance of the industry, particular operators, and technological trends to create key focus areas that change over time. Changes in focus areas are based on development plans, activities, audit plans, safety-critical activities, input from class societies, experience with operators as a whole and with individual operators, and new or revised regulations. PSA's current priority areas are

1. Assuring top management's role in managing major risks,
2. Conducting specific studies of technical and operational barriers (on the basis of risk and incidents),
3. Reducing risk to the external environment from subsea operations, and
4. Focusing on occupational risks to specific groups of people, such as sand blasters.

PSA also plans to change its system and program periodically as the sources of risk change. The Macondo well accident led PSA, like many regulators, to conduct a detailed investigation of the blowout's causes and the industry's response (PSA Norway 2011a, 2011b). Particular interest was paid to the question, "Could this happen in Norway?" and to what changes should be made in how PSA manages safety. The formal conclusion was that there was no reason to revise the system and no need for major overhaul, but that there were issues that the industry needed to address in light of the Macondo accident, including lack of understanding of risks, lack of supervision, and failure to follow procedures.

PSA posed this question to the industry: "Do you think you can operate safely without a capping and containment system?" The industry response was no. PSA also asked the industry, "Do we need better organization of emergency response?" The industry response was yes. So while some changes were and are being made in Norway in response to the Macondo accident, PSA has not seen a need to change its basic approach to ensuring that adequate management systems are in place on the Norwegian Continental Shelf.[21]

CENTER FOR OFFSHORE SAFETY: A SELF-POLICING SAFETY ORGANIZATION

Like the nuclear power industry in 1979—in the immediate aftermath of the Three Mile Island accident—the nation's oil and gas industry needs now to embrace the potential for an industry safety institute to supplement government oversight of industry operations.
— National Commission on the BP *Deepwater Horizon* Oil Spill and Offshore Drilling[22]

The National Commission on the BP *Deepwater Horizon* Oil Spill and Offshore Drilling (the presidential commission) recommended that a self-policing safety institute be set up by and for the companies working offshore. This proposal recognizes that although government regulators are not likely to achieve the technical safety expertise of private industry, the nation must have a high level of assurance that operations on the Outer Continental Shelf (OCS) are as safe as possible. In this regard, the commission thought that the Institute of Nuclear Power Operations, which was set up by the nuclear power industry after the accident at Three Mile Island, was the desirable model for the U.S. offshore oil and gas industry, although the commission recognized that the number of nuclear facilities that the Institute of Nuclear Power Operations oversees is far smaller than the number of OCS facilities in U.S. waters.

In March 2011, largely in response to the presidential commission's recommendation, and after some internal deliberation, the industry set up COS, the self-described mission of which is to promote the highest level of safety for offshore drilling, completions, and operations through effective leadership, communication, teamwork, use of disciplined SMSs,

---

[21] M. Ognedal, PSA Norway. Presentation to the committee, Houston, Texas, October 19, 2011.
[22] *Deep Water: The Gulf Oil Disaster and the Future of Offshore Drilling,* 2011, p. 241.

and independent third-party auditing and certification.[23, 24] The Committee for Analysis of Causes of the *Deepwater Horizon* Explosion, Fire, and Oil Spill to Identify Measures to Prevent Similar Accidents in the Future endorsed the concept of a center for offshore safety to train, monitor, and certify (license) offshore oil and gas personnel, stating,

> This center has the potential to engage the CEOs of oil and gas companies, drilling contractors, and service companies in risk management; set standards for training and certification; develop accreditation systems for industry training programs; and facilitate industry participation in safety audits and inspections." (NAE-NRC 2011, p. 121)

According to the COS website[25] and other information provided to the committee, a key operational feature of the center will be a process for independent validation of SEMS programs, with API RP 75 as the basis for the auditing program. The process will encompass audit protocols with metrics for the new SEMS regulation, third-party audits, and accreditation and certification of audit service providers. A major objective of COS is to have BSEE embrace the center's accredited third-party audits as an effective means of complying with regulations and improving industry performance.

Although COS is still in the process of being established, some discernible progress is being made. For example, the COS office just opened in Houston, and its governing board is virtually in place. When fully appointed, the board will have a maximum of 24 members, including an executive director. The allocation of seats on the board is intended to achieve a balance between producer–operator members and drilling contractor and service supply companies. Membership is open to all companies that operate, drill, or complete wells or provide support services to deepwater drilling, completions, and operations. A company does not have to be a member of API to be a member of the center; however, all API members that work on the OCS must become members of COS.

---

[23] J. Toellner, ExxonMobil. Presentation to the committee, Houston, Texas, October 19, 2011.
[24] C. Williams, Shell Energy Resources Co. Presentation to the committee, Washington, D.C., August 31, 2011.
[25] See http://www.centerforoffshoresafety.org/main.html.

COS is organized within API, and the COS governing board was established by the API executive committee. The chairman of the board is nominated by the API Upstream Committee and approved by the API executive committee for a term of 3 years. According to the COS website, the center is "organized within API to leverage the existing resources and experience embodied in the long established API certification and standards group."[26]

The integration of this nascent self-policing safety organization and API presents a significant credibility problem for COS and was a major concern of the presidential commission, which strongly urged that the new safety institute be completely separate from API. API, known for representing virtually all aspects of the oil and gas industry, is a consensus organization that generally settles on that to which a broad majority of interested member companies will agree. It is an organization that has many missions and objectives, including lobbying and policy advocacy. The committee, however, believes COS should have only one function—safety, both of the personnel working on offshore facilities and of the surrounding coastal and marine environment.

Nevertheless, it was probably inevitable that the initial offshore safety organization would be set up by API. API's standards and certification unit, which is the nonadvocacy part of the organization, does so much technical work that it would probably have been difficult to get support to create a parallel and completely independent institute with enough leadership commitment in time and money.

Only time will tell whether COS can be an effective, independent force for safety. It is helpful that COS is now based in Houston rather than Washington, D.C., and that it was formed by the standards committees of API rather than the policy advocacy arm. The presidential commission recommended that the new safety institute be established by the companies, and, notwithstanding the commission's clear concerns about credibility, an API relationship was the industry's decision.

The COS leadership will need to demonstrate over time that it can set a direction independent from API. COS must show that the SEMS programs of the companies working offshore are deserving of the nation's

---

[26] See http://www.centerforoffshoresafety.org/governance.html.

trust and confidence. This is a serious challenge, but one that the industry must succeed in meeting if it is going to convince the nation, including the government officials in BSEE who regulate the industry, that the safety mission of the offshore energy companies will not be compromised.

SUMMARY

It seems clear from the experiences of the regulatory agencies discussed in this chapter, especially CSLC and PSA Norway, that agencies with charges similar to those of BSEE and many years of experience in overseeing the SMS programs of offshore operators have found that issuing incident of noncompliance notices against a checklist of yes or no requirements[27] tends to lead to a culture of compliance rather than a culture of safety. Instead, these agencies have migrated toward a system that

- Audits operations with a qualified team of auditors,
- Discusses with personnel at different levels of the operation the way in which the elements of the SMS are actually being used,
- Feeds the results back to the top management of the operating companies, and
- Monitors for continuous improvement.

These agencies have found that engagement with the industry is more productive than punishment, although they maintain the threat of punishment if needed. Each of these agencies has developed a program and established regulations to assure the compliance of the specific activities and cultures of the industries under its purview. Each of these agencies has uniquely tailored its regulatory role so as to assure the compliance of the specific activities and cultures of the industries under its purview. In doing so, however, each has been moving from prescriptive regulations to a goal-oriented or risk-based approach of regulatory oversight in order to better promote continuous improvement in safety.

Even before the Macondo well blowout, MMS had undertaken efforts to change regulations for the offshore oil and gas industry, but the blow-

---

[27] For example, "Have written operating procedures been developed and implemented which include the job title and reporting relationship of the person(s) responsible for each of the facility's operating areas?" [30 CFR, Part 250, Section 250.1913(a)].

out was the catalyst for swift and sweeping regulatory changes, including the restructuring of MMS.[28] In response to the Macondo accident and these regulatory changes, the offshore oil and gas industry established COS, whose mission is to promote the highest level of safety through effective leadership, communication, teamwork, use of SMSs, and process auditing and certification. Although still in the process of being established, COS has the potential to be of great value to the industry.

---

[28] The history of the restructuring of MMS is discussed in both the preface and Chapter 1.

5

Role of the Bureau of Safety and Environmental Enforcement in Evaluating Safety and Environmental Management Systems Programs

The mission of the Bureau of Safety and Environmental Enforcement (BSEE) is "to promote safety, protect the environment, and conserve resources offshore through vigorous regulatory oversight and enforcement." One of its key functions is to develop "standards and regulations to enhance operational safety and environmental protection for the exploration and development of offshore oil and natural gas on the U.S. Outer Continental Shelf (OCS)."[1] In fulfilling this function, the Bureau of Ocean Energy Management, Regulation and Enforcement (BOEMRE, now BSEE) issued the Safety and Environmental Management Systems (SEMS) regulation, which requires operators and their contractors to establish and maintain a SEMS program (BOEMRE 2010). This chapter discusses considerations related to inspections and audits.

INSPECTIONS

As generally defined, an inspection is an organized examination or formal evaluation exercise. An inspection involves applying measurements, tests, and gauges to certain characteristics with regard to an object or activity. The results are usually compared with specified minimum requirements and standards for determining whether the item or activity meets these targets. Each operation, personnel action, system, subsystem, and component of a regulated entity under the jurisdiction of these requirements (i.e., regulations) is subject to this practice.

---

[1] See http://www.bsee.gov/About-BSEE/index.aspx.

Inspectors use their training, education, experience, and understanding of the intent of the particular regulation to determine compliance. To be ultimately successful, inspectors must be familiar with the equipment they inspect and the safe operating practices necessary to complete the task at hand. The regulations articulate the minimum standards necessary for compliance and, in doing so, limit inspectors to being able to require only these minimums.

Among its many provisions, the Outer Continental Shelf Lands Act contains several safety-related directives, including one that requires that each facility on the OCS be subject to annual scheduled inspections of all safety equipment designed to prevent or ameliorate blowouts, fires, spillages, or other major accidents. Additionally, the law includes a requirement for periodic onsite inspections, without advance notice to the operator of any facility, to ensure compliance with such environmental or safety regulations. Thus, there is a requirement predating SEMS that BSEE continue to inspect offshore installations. A memorandum of understanding between the U.S. Coast Guard (USCG) and the Minerals Management Service (MMS) was enacted in 2004 to minimize duplication between agencies and to promote consistent regulation of facilities and operations on the OCS (MMS-USCG 2004b). To date, portions of this memorandum of understanding have been clarified or revised, or both, in four follow-up memoranda of agreement (MMS-USCG 2004a, 2006a, 2006b, 2008).

The 1990 Marine Board report *Alternatives for Inspecting Outer Continental Shelf Operations* states,

> There is a strong sentiment in the industry, on the part of offshore operators and employees, as well as MMS employees, that the regular presence of MMS personnel has positive benefits on safety which should not be foregone. (NRC 1990, p. 72)

and

> The presence of government inspectors on the OCS is important for conveying a sense of oversight and for providing impetus to marginal and inexperienced operators to meet federal safety standards. (NRC 1990, p. 81)

Members of the Committee on the Effectiveness of Safety and Environmental Management Systems for Outer Continental Shelf Oil and Gas Operations (the committee) heard similar statements from both operator and MMS personnel on a trip to an installation off the Pacific Coast in March 2010. Thus, any system for evaluating the effectiveness of SEMS will need to include a continuing presence of BSEE inspectors on offshore installations.

The executive summary of the Marine Board report states,

> A final point made by the committee—and it is a crucial one—relates to attitudes. In enterprises that are subject to inspection by government or other authorities, the operators of the enterprise often gradually drift to the point of view that the responsibility for safety lies with the government and the inspectors. An attitude develops that the operator's responsibility and objective is simply to pass the inspection, an attitude the committee refers to as a "compliance mentality." It is especially likely to develop when inspections are based on a routine checklist approach.
>
> The committee emphasizes its belief that *compliance does not equal safety*. Thus, although it is certainly desirable to have checklists to guide inspectors, it is important for MMS to ensure that operators do not sink into a compliance mentality. To reiterate: in practice and in law, the operators bear the primary responsibility for safety. The MMS, for its part, is responsible for using the best and most efficient means it can devise to motivate operators to meet that responsibility. (NRC 1990, p. 5)

The report goes on to state

> A key question is, 'What is the actual relationship between inspection and the safety of offshore platforms?' It is a truism that inspection contributes positively to safety, but it is widely accepted by safety professionals that too much inspection, the wrong kind of inspection, or the wrong attitude about inspection can detract from safety." (NRC 1990, p. 39)

The committee endorses these sentiments. BSEE has a role in helping the industry develop the culture of safety that the government, the industry, and the public want. The manner in which BSEE evaluates the effectiveness of SEMS can help or hinder this effort, and BSEE needs to take this into account when determining its role.

AUDITS

In establishing its role, BSEE must take care to consider appropriately the role of the operating companies. The initial SEMS rule, issued in October 2010, became effective on November 15, 2011 (30 CFR 250,

Subpart S). Under Sections 250.1920 and 250.1921, SEMS currently requires the operator to conduct an audit program using either its own qualified employees (i.e., an internal audit) or a qualified third party, as follows:

§ 250.1920 What are the auditing requirements for my SEMS program?

(a) You must have your SEMS program audited by either an independent third-party or your designated and qualified personnel according to the requirements of this subpart and API [American Petroleum Institute] RP [Recommended Practice] 75, Section 12 (incorporated by reference as specified in § 250.198) within 2 years of the initial implementation of the SEMS program and at least once every 3 years thereafter. The audit must be a comprehensive audit of all thirteen elements of your SEMS program to evaluate compliance with the requirements of this subpart and API RP 75 to identify areas in which safety and environmental performance needs to be improved.

(b) Your audit plan and procedures must meet or exceed all of the recommendations included in API RP 75 section 12 (incorporated by reference as specified in § 250.198) and include information on how you addressed those recommendations. You must specifically address the following items:

(1) Section 12.1 General.
(2) Section 12.2 Scope.
(3) Section 12.3 Audit Coverage.
(4) Section 12.4 Audit Plan. You must submit your written Audit Plan to BSEE at least 30 days before the audit. BSEE reserves the right to modify the list of facilities that you propose to audit.
(5) Section 12.5 Audit Frequency, except your audit interval must not exceed 3 years after the 2 year time period for the first audit.
(6) Section 12.6 Audit Team. The audit that you submit to BSEE must be conducted by either an independent third party or your designated and qualified personnel. The independent third party or your designated and qualified personnel must meet the requirements in § 250.1926.

(c) You must require your auditor (independent third party or your designated and qualified personnel) to submit an audit report of the findings and conclusions of the audit to BSEE within 30 days of the audit completion date. The report must outline the results of the audit, including deficiencies identified.

(d) You must provide the BSEE a copy of your plan for addressing the deficiencies identified in your audit within 30 days of completion of the audit.

Your plan must address the following:

(1) A proposed schedule to correct the deficiencies identified in the audit. BSEE will notify you within 14 days of receipt of your plan if your proposed schedule is not acceptable.

(2) The person responsible for correcting each identified deficiency, including their job title.

(e) BSEE may verify that you undertook the corrective actions and that these actions effectively address the audit findings.

Thus an audit and report are required every 3 years after the initial audit. The audit must address all 13 elements of SEMS—17 elements if the SEMS II notice of proposed rule making (BOEMRE 2011a) is adopted (see Table 4-1)—and BSEE must preapprove the audit plan.

The number of installations that must be covered by each audit is not specified in 30 CFR 250, but reference is made to API RP 75, which requires that each audit include coverage of at least 15 percent of the operator's facilities. API RP 75, Section 12.3, "Audit Coverage," states,

> When selecting facilities to audit, consideration should be given to common features (e.g., field supervisors, regulatory districts, facility design, systems and equipment, office management, etc.) to obtain a cross-section of practices for the facilities operated. The testing system of the audit need not be applied to each facility; rather, interviews and inspections should be conducted at fields that differ significantly (e.g., oil vs. dry gas). This should include a number of facilities sufficient to evaluate management's commitment to items a, b, and c in 12.2. During each audit, at least fifteen percent (15%) of the facilities operated, with a minimum of one facility, should be audited. The facilities included in the audit should not be the same as those included in the previous audit. When sufficient deficiencies are identified in the effectiveness of any safety and environmental management program elements, the test sample size shall be expanded for that program element. (API 2004, p. 25)

Thus, every 3 years, each operator must audit at least one of its installations. Operators with multiple installations need only audit a representative sample of 15 percent of the installations.

BSEE also reserves the right to conduct audits of its own or to require an operator to have a third party conduct an audit, as specified in 30 CFR 250.1925:

§ 250.1925 May BSEE direct me to conduct additional audits?

(a) If BSEE identifies safety or non-compliance concerns based on the results of our inspections and evaluations, or as a result of an event, BSEE may direct you to have an independent third-party audit of your SEMS program, in addition to the regular audit required by § 250.1920, or BSEE may conduct an audit.

(1) If BSEE directs you to have an independent third-party audit,

(i) You are responsible for all of the costs associated with the audit, and

(ii) The independent third-party audit must meet the requirements of § 250.1920 of this part and you must ensure that the independent third party submits the findings and conclusions of a BSEE-directed audit according to the requirements in § 250.1920 to BSEE within 30 days after the audit is completed.

(2) If BSEE conducts the audit, BSEE will provide a report of the findings and conclusions within 30 days of the audit.

(b) Findings from these audits may result in enforcement actions as identified in § 250.1927.

(c) You must provide the BSEE a copy of your plan for addressing the deficiencies identified in the BSEE-directed audit within 30 days of completion of the audit as required in § 250.1920.

Auditor Qualifications

SEMS audits span a wide range of disciplines and require auditors who are suitably qualified and trained in the technical skills involved in offshore safety and environmental issues as well as in the audit function. Auditing can be performed by organizations or individuals, and both should be competent as well as independent. Consideration will need to be given to the various tasks associated with the audit function as well as to the qualifications of the individuals authorized to perform those tasks.

Section 250.1926 of the initial SEMS rule describes the required minimum qualifications of the individual or organization conducting the audit and requires BSEE to approve the qualifications of each auditor:

§ 250.1926 What qualifications must an independent third party or my designated and qualified personnel meet?

(a) You must either choose an independent third-party or your designated and qualified personnel to audit your SEMS program. You must take into account the following qualifications when selecting the third-party or your designated and qualified personnel:

    (1) Previous education and experience with SEMS, or similar management related programs.

    (2) Technical capabilities of the individual or organization for the specific project.

    (3) Ability to perform the independent third-party functions for the specific project considering current commitments.

    (4) Previous experience with BSEE regulatory requirements and procedures.

    (5) Previous education and experience to comprehend and evaluate how the company's offshore activities, raw materials, production methods and equipment, products, byproducts, and business management systems may impact health and safety performance in the workplace.

(b) You must have procedures to avoid conflicts of interest related to the development of your SEMS program and the independent third party auditor and your designated and qualified personnel.

(c) BSEE may evaluate the qualifications of the independent third parties or your designated and qualified personnel. This may include an audit of documents and procedures or interviews. BSEE may disallow audits by a specific independent third-party or your designated and qualified personnel if they do not meet the criteria of this section.

These qualifications, which are under consideration for modification in the SEMS II notice of proposed rulemaking (BOEMRE 2011a), reflect the basic high-level qualifications needed of auditors:

- Education and previous experience with SEMS or similar management-related programs;
- Previous experience with BSEE regulatory requirements and procedures; and
- Educational background and previous experience relevant to understanding and evaluating how the operator's offshore activities, raw materials, production methods and equipment, products, by-products, and business management systems may affect health and safety performance in the workplace.

In addition to specifying these qualifications, SEMS II addresses the independence of the auditor with the following requirements (BOEMRE 2011a):

- The operator must provide a document signed by its management that states that the independent third-party auditor is not owned or controlled by, or otherwise affiliated with, the operator's company.
- The operator must have procedures for avoiding conflicts of interest related to the development of its SEMS program and to the independent third-party auditor. If an independent third party developed or maintains the SEMS program, then that person or its subsidiaries cannot audit the program.

What Makes a Good Auditor?
The SEMS requirements summarized above should be considered minimum requirements. The operator should view the audit not only as an opportunity to confirm both compliance with the regulations and the effectiveness of its SEMS program but, more important, as an opportunity to have a positive impact on the organization and further enhance its safety culture. The best auditors work with the organizations they are auditing by encouraging industry best practices to promote continuous improvement at all levels of the organization. They are familiar with the operations and responsibilities of the facility being audited and are sometimes recognized as being interested in improving the performance of the safety management system rather than as being an enforcer or punisher.

Therefore, auditors must have special skills that are achieved through education, training, and experience. Numerous existing auditing protocols and qualification requirements are available as examples for BSEE's SEMS auditors. A majority of the organizations with programs similar to SEMS that are discussed in Chapter 4, such as Petroleum Safety Authority Norway, the United Kingdom Health and Safety Executive, and the Occupational Safety and Health Administration, have similar auditing protocols and qualifications, and details can be found in the associated references for each.

Auditor Competence

ISO 9001, *Quality Management Systems* (ISO 2008), is a widely accepted international standard that is in use in more than 1 million organizations worldwide. Many of the offshore oil and gas operators subject to SEMS use ISO 9001 as the basis for their quality management system. ISO 9001 requires auditing and references a separate document, ISO 19011, *Guidelines for Quality and/or Environmental Management Systems Auditing* (ISO 2002; BSI 2012), as a basis for the audit process. ISO 19011 covers managing audits, audit activities, preparing for and conducting audits, preparing audit reports, and conducting follow-up activities.

ISO 19011 also provides considerable guidance on the competence and quality of good auditors. It defines competence as "ability to apply knowledge and skills to achieve intended results" (BSI 2012, p. 3).[2] The competence of auditors involves a combination of characteristics, the key aspects of which are knowledge and skills, education, work experience, auditor training, audit experience, and personal attributes (Figure 5-1).

Knowledge and Skills   Auditors must have a combination of generic as well as SEMS-specific knowledge and skills that pertain to the safety and environmental aspects of offshore operations. Generic knowledge and skills consist of basic auditing principles and techniques, including the ability to plan and execute the audit effectively. This type of knowledge is applicable to any type of audit, including SEMS. Knowledge and skills specific to the safety and environmental aspects of offshore operations include knowledge of the BSEE SEMS standard, including the related science, technology, processes, and terminology, as well as the interface between systems and human activities.

Education, Work Experience, Auditor Training, and Audit Experience
Auditor education, experience, and training should complement each other. Education should be sufficient in the technical skills associated

---

[2] Permission to reproduce extracts from *Guidelines for Auditing Management Systems (ISO 19011:2011)* is granted by BSI. British Standards can be obtained in PDF or hard copy formats from the BSI online shop: www.bsigroup.com/Shop or by contacting BSI Customer Services for hardcopies only: tel: +44 (0)20 8996 9001, e-mail: cservices@bsigroup.com.

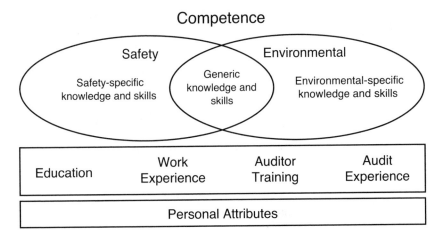

FIGURE 5-1 Concept of competence and auditor qualifications related to SEMS (modified by committee from ISO 19011 to apply to SEMS).

with offshore safety, the environment, and auditing processes. ISO 19011 (BSI 2012, p. 28)[3] states

> Auditor knowledge and skills can be acquired using a combination of the following:
>
> - formal education/training and experience that contribute to the development of knowledge and skills in the management system discipline and sector the auditor intends to audit;
> - training programmes that cover generic auditor knowledge and skills;
> - experience in a relevant technical, managerial or professional position involving the exercise of judgement, decision making, problem solving and communication with managers, professionals, peers, customers and other interested parties;
> - audit experience acquired under the supervision of an auditor in the same discipline.

Methods for evaluating the auditor for competence and for maintaining and improving auditor competence are also described.

---

[3] Permission to reproduce extracts from *Guidelines for Auditing Management Systems (ISO 19011:2011)* is granted by BSI. British Standards can be obtained in PDF or hard copy formats from the BSI online shop: www.bsigroup.com/Shop or by contacting BSI Customer Services for hardcopies only: tel: +44 (0)20 8996 9001, e-mail: cservices@bsigroup.com.

Personal Attributes As noted several times in this report, the success (or failure) of an audit greatly depends on the auditor(s) involved. To be effective, the auditor(s) must have not only the proper technical knowledge, skills, and education, but also good interpersonal skills. An audit is an emotional event in which an organization and its employees are examined, and the auditor's approach to this process is highly important. A person may have the proper technical knowledge and skills but a poor personal approach that alienates. ISO 19011 (BSI 2012, pp. 25–26)[4] recommends that

> Auditors should exhibit professional behaviour during the performance of audit activities, including being:
>
> - ethical, i.e., fair, truthful, sincere, honest and discreet;
> - open-minded, i.e., willing to consider alternative ideas or points of view;
> - diplomatic, i.e., tactful in dealing with people;
> - observant, i.e., actively observing physical surroundings and activities;
> - perceptive, i.e., aware of and able to understand situations;
> - versatile, i.e., able to readily adapt to different situations;
> - tenacious, i.e., persistent and focused on achieving objectives;
> - decisive, i.e., able to reach timely conclusions based on logical reasoning and analysis;
> - self-reliant, i.e., able to act and function independently whilst interacting effectively with others;
> - [able to act] with fortitude, i.e., able to act responsibly and ethically, even though these actions may not always be popular and may sometimes result in disagreement or confrontation;
> - open to improvement, i.e., willing to learn from situations, and striving for better audit results;
> - culturally sensitive, i.e., observant and respectful to the culture of the auditee;
> - collaborative, i.e., effectively interacting with others, including audit team members and the auditee's personnel.

---

[4] Permission to reproduce extracts from *Guidelines for Auditing Management Systems (ISO 19011:2011)* is granted by BSI. British Standards can be obtained in PDF or hard copy formats from the BSI online shop: www.bsigroup.com/Shop or by contacting BSI Customer Services for hardcopies only: tel: +44 (0)20 8996 9001, e-mail: cservices@bsigroup.com.

Basic Underlying Principles of Auditing

Auditing is characterized not only by the competency issues discussed above, but by reliance on several core principles that must carry down directly to the auditor. These principles revolve around ethics, professionalism, and independence and provide the basis for developing audit conclusions that are unbiased and pertinent. They also provide the basis for repeatability (i.e., no matter who performs the audit, the findings would be similar for similar timing, situations, and circumstances). ISO 19011 (BSI 2012, pp. 4–5)[5] summarizes these principles as follows:

a) **Integrity:** the foundation of professionalism
   Auditors and the person managing an audit programme should:
   – perform their work with honesty, diligence, and responsibility;
   – observe and comply with any applicable legal requirements;
   – demonstrate their competence while performing their work;
   – perform their work in an impartial manner, i.e., remain fair and unbiased in all their dealings;
   – be sensitive to any influences that may be exerted on their judgement while carrying out an audit.
b) **Fair presentation:** the obligation to report truthfully and accurately
   Audit findings, audit conclusions and audit reports should reflect truthfully and accurately the audit activities. Significant obstacles encountered during the audit and unresolved diverging opinions between the audit team and the auditee should be reported. The communication should be truthful, accurate, objective, timely, clear and complete.
c) **Due professional care:** the application of diligence and judgement in auditing
   Auditors should exercise due care in accordance with the importance of the task they perform and the confidence placed in them by the audit client and other interested parties. An important factor in carrying out their work with due professional care is having the ability to make reasoned judgements in all audit situations.

---

[5] Permission to reproduce extracts from *Guidelines for Auditing Management Systems (ISO 19011:2011)* is granted by BSI. British Standards can be obtained in PDF or hard copy formats from the BSI online shop: www.bsigroup.com/Shop or by contacting BSI Customer Services for hardcopies only: tel: +44 (0)20 8996 9001, e-mail: cservices@bsigroup.com.

d) **Confidentiality:** security of information
Auditors should exercise discretion in the use and protection of information acquired in the course of their duties. Audit information should not be used inappropriately for personal gain by the auditor or the audit client, or in a manner detrimental to the legitimate interests of the auditee. This concept includes the proper handling of sensitive or confidential information.

e) **Independence:** the basis for the impartiality of the audit and objectivity of the audit conclusions
Auditors should be independent of the activity being audited wherever practicable, and should in all cases act in a manner that is free from bias and conflict of interest. For internal audits, auditors should be independent from the operating managers of the function being audited. Auditors should maintain objectivity throughout the audit process to ensure that the audit findings and conclusions are based only on the audit evidence. For small organizations, it may not be possible for internal auditors to be fully independent of the activity being audited, but every effort should be made to remove bias and encourage objectivity.

f) **Evidence-based approach:** the rational method for reaching reliable and reproducible audit conclusions in a systematic audit process
Audit evidence should be verifiable. It will in general be based on samples of the information available, since an audit is conducted during a finite period of time and with finite resources. An appropriate use of sampling should be applied, since this is closely related to the confidence that can be placed in the audit conclusions.

Training and Certification

Training programs allow individuals to become familiar with audit requirements. Structuring training programs around the elements of SEMS will allow a focus on qualifications that pertain to specific elements, so that auditors can be authorized to perform particular functions. A SEMS audit team could then be composed of individuals with different levels of competence and authorization.

Training can be conducted either in-house or externally, and there are companies developing training specific to SEMS. Training courses, whether given internally or externally, should be tested and independently certified.

As described in Chapter 4, the Center for Offshore Safety (COS) is an industry group that, in association with API, is currently developing a third-party certification and auditor process specifically for SEMS. The COS plan is to certify audit service providers to conduct SEMS audits using a protocol developed by COS that satisfies both BSEE and industry requirements. Because COS may be involved in many SEMS audits, it plans to collect what it learns about best practices associated with the audit process and then share this information with the industry. The accreditation process and qualification requirements for COS auditors were still under development at the writing of this report, but it is anticipated that these will include many of the characteristics discussed in this chapter.

ENSURING EFFECTIVENESS

The role that BSEE will play in ensuring that a SEMS program is in place and operating properly is described in 30 CFR 250.1924:

§ 250.1924 How will BSEE determine if my SEMS program is effective?

(a) BSEE or its authorized representative may evaluate or visit your facility to determine whether your SEMS program is in place, addresses all required elements, and is effective in protecting the safety and health of workers, the environment, and preventing incidents. BSEE or its authorized representative may evaluate your SEMS program, including documentation of contractors, independent third parties, your designated and qualified personnel, and audit reports, to assess your SEMS program. These evaluations or visits may be random or based upon the OCS lease operator's or contractor's performance.

(b) For the evaluations, you must make the following available to BSEE upon request:

(1) Your SEMS program;

(2) The qualifications of your independent third-party or your designated and qualified personnel;

(3) The SEMS audits conducted of your program;

(4) Documents or information relevant to whether you have addressed and corrected the deficiencies of your audit; and

(5) Other relevant documents or information.

(c) During the site visit BSEE may verify that:
   (1) Personnel are following your SEMS program;
   (2) You can explain and demonstrate the procedures and policies included in your SEMS program; and
   (3) You can produce evidence to support the implementation of your SEMS program.
(d) Representatives from BSEE may observe or participate in your SEMS audit. You must notify the BSEE at least 30-days prior to conducting your audit as required in § 250.1920, so that BSEE may make arrangements to observe or participate in the audit.

The SEMS regulation states that the agency will take certain enforcement actions if it finds the SEMS program or its audits to be out of compliance (30 CFR 250.1927):

§ 250.1927 What happens if BSEE finds shortcomings in my SEMS program?

If BSEE determines that your SEMS program is not in compliance with this subpart we may initiate one or more of the following enforcement actions:
(a) Issue an Incident(s) of Noncompliance;
(b) Assess civil penalties; or
(c) Initiate probationary or disqualification procedures from serving as an OCS operator.

Thus, the current role of the operator is to establish a SEMS program and conduct specified internal or third-party audits according to a plan approved by BSEE.

BSEE's current role is to either visit facilities themselves or arrange for third parties to do so on behalf of BSEE to inspect for compliance, approve all operator audit plans and individual auditors, review the results of all audits, and issue incident of noncompliance (INC) notices or other forms of punishment for noncompliance. It is unclear whether the intent is to issue punishment for deficiencies found in an operator's audit(s) as well as for deficiencies found in BSEE-arranged inspections and audits.

In discussions with BSEE, the committee was told that current BSEE division inspectors would have checklists and be expected to issue INCs for any deficiencies in an operator's SEMS program that the inspectors observed during their inspections;[6] however, it is not expected that these BSEE employees will be trained and qualified as SEMS auditors. Rather, the regions will hire and train a cadre of qualified auditors who will be able to review audit plans and reports and conduct audits on behalf of BSEE.

The committee recognizes that it will likely take some time for proper training and development programs to be implemented by BSEE, as well as time to locate and hire appropriate personnel. This likely will not be easy or quick, but will be necessary in order to move beyond following a checklist and toward promoting a culture of safety. One necessary, but likely not sufficient, step toward achieving the desired competence is requiring each BSEE auditor to fulfill certification requirements comparable to those needed by third-party auditors.

In a notice in the *Federal Register* on September 14, 2011, BOEMRE announced its intention to amend SEMS (30 CFR 250 Subpart S) as follows (BOEMRE 2011a):

(1) Procedures to authorize any and all employees on the facility to implement a Stop Work Authority (SWA) program when witnessing an activity that is regulated under BOEMRE jurisdiction that creates a threat of danger to an individual, property, and/or the environment;

(2) Clearly defined requirements establishing who has the ultimate authority on the facility for operational safety and decision making at any given time;

(3) A plan of action that shows how operator employees are involved in the implementation of the API's Recommended Practice for Development of a Safety [and] Environmental Management Program for Offshore Operations and Facilities (API RP 75), as incorporated by reference in the subpart S regulatory requirements in the October 15, 2010, final rule;

(4) Guidelines for reporting unsafe work conditions related to an operator's SEMS program, that provide all employees the right to report a possible safety or environmental violation(s) and to request a BOEMRE inspection of the facility if they believe there is a serious threat of danger or their employer is not following BOEMRE regulations;

---

[6] D. Slitor. BOEMRE Status Report. Presentation to the committee, Washington, D.C., August 31, 2011.

(5) Revisions that require operators with SEMS programs to engage independent third party auditors to conduct all audits of operators' SEMS programs and that the independent third party auditors must meet the criteria listed in Section 250.1926 of this proposed rule.
(6) Additional requirements for conducting a JSA [job safety analysis].

The revisions would prohibit operators from using their own qualified staff to conduct required SEMS audits. Instead, operators would be required to use BSEE-approved third-party auditors. The 1990 Marine Board study also looked at replacing MMS inspections with a requirement that MMS require operators to arrange for third-party inspections:

> It is hard to assess the impact this alternative might have on the safety consciousness of operators. With adequate precautions to obviate conflict-of-interest situations, there is no reason to believe that third-party inspectors would not carry out their duties so that the compliance element of the inspection process would be unchanged. This fact in itself, however, provides the operator with the same kind of shelter that he now has when he successfully "passes" an MMS inspection. Thus, the committee believes there will be a negligible impact on safety consciousness: the tendency toward a "compliance mentality" would not be corrected by this alternative. (NRC 1990, p. 72)

The committee, once again, endorses the concept that an evaluation system that maximizes the involvement of the operator in auditing and improving its SEMS procedures is preferable from the standpoint of moving from a compliance mentality to one of continuous development of a culture of safety.

6

Conclusions and Recommended Approach

Because government oversight alone cannot reduce risks to the full extent possible, the Bureau of Safety and Environmental Enforcement (BSEE) will need to look beyond its predecessor agencies' historical role of assuring compliance with prescriptive regulations and seize the current opportunity to design its role, at least partially, to encourage an atmosphere that helps the industry migrate from a compliance mentality to a culture of safety that includes compliance. To assist the agency in this endeavor, the Committee on the Effectiveness of Safety and Environmental Management Systems for Outer Continental Shelf Oil and Gas Operations (the committee) drew on the information obtained from presentations it received, site visits, published regulations, notices of proposed rulemaking, recommended practices, and previously published reports to develop the conclusions and recommended approach presented in this chapter.

CONCLUSIONS

Building Safety Culture Through Safety and Environmental Management Systems

Conclusion 1: If BSEE's goal is, as it should be, to encourage a culture of safety so that individuals know the safety aspects of their actions and are motivated to think about safety, then the agency will need to evolve an evaluation system for Safety and Environmental Management Systems (SEMS) that emphasizes the evaluation of attitudes and actions rather than documentation and paperwork. All of the elements of SEMS must be addressed, but it is much more important that those who are actually doing the work understand and implement SEMS than it is that SEMS documentation be verified with a checklist.

Conclusion 2: A SEMS program that contains all the elements laid out in the SEMS regulation is necessary but not sufficient for creating a culture of safety. An organization's safety culture will reduce risk; SEMS is but a means to that end.

a. A culture of safety must be supported throughout the organization—from the top to the bottom—to be effective.
b. A culture of safety only exists where the work occurs. If it does not actually drive the actions that people take, then it is only theoretical.

Merely following a strict interpretation of a minimal SEMS program will not guarantee safe operations offshore. An effective SEMS program cannot rely on checklist compliance; the program must become ingrained in the operation's management structure to be successful. The tenets of SEMS must be fully acknowledged and accepted by workers and be motivated from the top. Only then can an effective culture of safety be established and grow.

Conclusion 3: The operator "owns" the SEMS program and is responsible for ensuring that it is operating effectively. The operator's upper management is responsible and accountable for ensuring that a culture of safety exists. A safe operation is only possible when it is fully embraced by the organization. An organization cannot turn over the development and monitoring of its safety program to a third party and expect the program to be effective. Therefore, the ultimate responsibility for successful implementation of the SEMS program should reside with top management. If they do not take direct and complete ownership of the program, then safety will be relegated to a low status when difficult trade-offs need to be made.

Conclusion 4: To be effective, safety and environmental management must be a dynamic process that evolves with time and is reflected in the regulator's actions. Operations offshore are dynamic. Operating conditions, personnel, production requirements, and technologies are continually changing. Safety practices that were applicable during an earlier phase of operations may no longer be effective. Likewise, inspection and audit criteria will become outdated as new technologies are employed and new environments are explored. To be effective, safety procedures and the audits that verify their effectiveness should also be dynamic and informed by risk.

Conclusion 5: BSEE can encourage or hurt the development of a culture of safety by the way it measures and enforces SEMS. Forcing an operation to satisfy checklists that require specific forms of documentation and penalizing those operations that do not is likely to encourage a culture of compliance and discourage the development of a culture of safety.

Conclusion 6: A holistic combination of methods is necessary to evaluate the effectiveness and continuous improvement of an operator's SEMS program. Because of the diversity, complexity, and evolving nature of offshore oil and gas operations and the comprehensive nature of a fully implemented SEMS program, no single approach to inspections and audits will be sufficient to ensure a successful SEMS program. Both occupational safety and process, or system, safety need to be verified. A single one-dimensional snapshot of compliance will not provide the necessary insight upon which to construct a successful program.

Conclusion 7: All parties involved in a safety management program like SEMS share the common goal of safe operations. It is not possible to regulate a culture of safety by inspections or audits; that culture needs to come from within an organization. Regulators are an integral part of an organization's safety program, but they are limited as to what they can accomplish. The regulator's role should be to develop an approach for the inspection and auditing of increased safety rather than toward a path of compliance. It is important that BSEE inspectors and auditors do not direct or dictate specific action, because doing so would move responsibility from the operator to the regulator. In addition, it is important that BSEE strive to nurture safety culture within its own organizational system. As observed by the Organisation for Economic Co-operation and Development, "The nature of the relationship between the regulator and the operator can influence the operator's safety culture at a [facility] either positively or negatively. In promoting safety culture, a regulatory body should set a good example in its own performance" (OECD 1999, p.11).

Effective Inspections

Conclusion 8: The routine presence of competent BSEE inspectors on an offshore operator's facility is essential for verifying that the offshore oil and gas industry is generally complying with all of BSEE's regulations as well as providing a potential indication of compliance with SEMS.

Conclusion 9: To be effective at identifying problems inherent in an operator's safety culture, BSEE inspectors need to spend enough time on a facility to observe multiple activities.

Effective Audits

Conclusion 10: Audits, in and of themselves, are not sufficient to improve safety. For audit results to be effective, the operator needs to detect trends, identify deficiencies, take appropriate corrective action, and document the actions taken.

Conclusion 11: As part of its holistic approach to measuring effectiveness, BSEE is responsible for ensuring that the implementation of SEMS is audited; however, the primary responsibility for auditing the SEMS program rests with the operator.

a. A properly conducted, truly independent internal audit is potentially more effective than an independent third-party audit, as it reinforces ownership of the safety culture.
b. Some operators are too small to have sufficient staff to perform a truly independent internal audit and will need to use an independent third party to conduct the audit. Inclusion of at least one person from the operator's organization on the audit team will help cultivate management ownership of the audit. It would not be appropriate, however, to include a person directly involved in the day-to-day operation being audited.

A properly motivated, active in-house safety program can be the best vehicle for discovering and correcting unsafe practices. Under most circumstances, the personnel within an organization are the best equipped to identify both unsafe practices and feasible solutions. A third-party audit helps to ensure independence and can provide an outside perspective on operations, but at a cost in operator ownership. Almost by definition, a properly motivated and conducted internal safety audit requires buy-in from management; however, BSEE is responsible for ensuring that these internal audits are properly motivated and conducted. When resources do not permit an organization to conduct effective internal audits, third parties will need to be used.

Conclusion 12: BSEE is responsible for verifying that quality audits are carried out and acted on appropriately. Because of the comprehensive nature of the SEMS requirements, BSEE's oversight of internal and third-party audits needs to include a range of techniques, each of which focuses on a different aspect of an operation's safety system. These techniques are discussed below in the section on the committee's recommended approach.

Conclusion 13: Conducting a quality SEMS audit requires enough qualified personnel with sufficient time to spend on location. A quality SEMS audit requires

a. A mechanism for qualifying auditors,
b. An audit team with the skills required for the specific audit, and
c. An understanding of how those performing the work perceive SEMS, in addition to a review of SEMS documentation.

Conclusion 14: The skills and competencies required by inspectors are different from those required by auditors. To ensure that operators have established and are maintaining viable SEMS programs, BSEE needs to employ personnel skilled and well trained in two different areas: inspecting and auditing. Inspections that rely primarily on checklist compliance require inspectors with firsthand knowledge of equipment characteristics and procedures. In addition to this knowledge, the personnel charged with auditing a SEMS program need a higher-level knowledge of operations, a detailed knowledge of SEMS, a thorough understanding of how operating procedures and safety performance are related, and training in performing audits.

Conclusion 15: Because BSEE will have access to all SEMS programs, including audit and follow-up reports and its own inspection and audit reports, the agency is in the best position to gather and analyze this data to identify best practices and common trends (good and bad), and disseminate this information to the regulated community in a timely manner. BSEE auditing of SEMS programs will encompass a wide range of different operators that are using a variety of approaches to SEMS. Some of these programs may be more effective than others and will provide valuable lessons that can be used to improve SEMS for the entire industry.

RECOMMENDED APPROACH

BSEE should establish a holistic combination of methods necessary to ensure the effectiveness and continuous improvement of SEMS programs. BSEE should establish a system that employs a combination of compliance inspections, audits, key performance indicators (KPIs), and a whistleblower program to ensure that SEMS programs are adequate, in place, and operating effectively and in a manner that promotes a culture of safety among the operators. The committee did not reject outright any of the approaches presented in Chapter 3 as being of no value, and BSEE could employ a combination of all the methods presented in Chapter 3. However, on the basis of the experience of representatives from comparable regulatory agencies (see Chapter 4) and of the committee members' personal experience and expertise, the committee selected a combination of those approaches for which it believed some evidence of success was available. These approaches, the committee believed, would result in the most effective evaluation with the resources available. The technology and data necessary to use the SEMS monitoring sensors and calculation of risk methods described in Chapter 3 are not available at this time.

The following sections of this chapter provide specific details of the recommended approach. Specific recommendations are given in boldface. The first section describes the inspection criteria and procedures that BSEE could use to verify compliance with specific regulations and to uncover obvious deficiencies in the implementation of the operator's SEMS program. The second section discusses audits and describes a process by which BSEE could ensure that operator-initiated audits are complete, accurate, and effective (i.e., an audit of the audit process). The final sections summarize recommendations for a KPI program and a whistleblower program. Together, these methods constitute a holistic system that BSEE should employ to evaluate the effectiveness of SEMS.

BSEE SEMS Inspections

BSEE should continue its current program of ensuring compliance with specific regulations. However, **an inspection of SEMS (scheduled or otherwise) should not be focused solely on what is not in compliance; rather, the inspection should attempt to obtain a holistic view of the facility's safety culture.** Does the operating company empower its personnel to take corrective action? Does it provide the resources necessary to do so? Are facility personnel only focused on the items they know an inspector typically will look at, and do they ignore the rest? A well-trained evaluator should be able to look beyond the black-and-white regulation and identify operators in marginal compliance, in order to guide them into a more complete state of compliance.

BSEE inspectors have been trained to measure compliance with a standard or prescriptive checklist and without further training are not equipped to measure the effectiveness of individual SEMS programs. It may not be practical to expect the current BSEE inspection force to make subjective decisions as to whether a SEMS program is working correctly so that it meets the intent of the SEMS regulation and helps create a culture of safety. Current BSEE inspectors will have a tendency to issue incident of noncompliance (INC) notices for deviations of documentation from a checklist, and such deviations may or may not be important in meeting the intent of SEMS. In turn, the issuing of INCs may focus operator attention on compliance in the way documentation is written rather than on establishing a culture that actually promotes safety.

Most injuries and blowouts on offshore oil and gas facilities are not usually caused by mechanical failures identified by INCs (NRC 1990). **Thus, to the extent possible, BSEE should train inspectors to employ other options in addition to using prescriptive checklists and issuing**

INCs. **Ideally, BSEE inspectors should look beyond the written regulation to identify operators in marginal compliance and guide them into a more complete state of compliance.** Although it may be difficult for BSEE inspectors to identify operators in marginal compliance when it comes to assessing the adequacy of a SEMS program, it is not unreasonable to expect them to make overall observations, which, in turn, could help focus BSEE-initiated SEMS audits.

During presentations and site visits[1] the committee was told that, in some cases, BSEE inspectors visiting an offshore installation might be able to spot problems inherent in an operator's safety culture by noticing obvious safety issues such as loose handrails, corroded walkways, or staff not wearing the appropriate personal protective gear. Other situations that indicate problems in the safety culture might be much harder to notice and require an in-depth investigation of safety-related approaches and practices, not only at the installation, but also in the operator's overall operations, both offshore and onshore.

BSEE inspectors need to spend enough time on a facility to observe multiple activities. To avoid the appearance of a conflict of interest, BSEE inspectors are not generally permitted to travel on operator-furnished helicopters, eat food provided by the operator while on the offshore facility, or stay overnight in operator-furnished quarters.[2] Every other offshore regulatory regime the committee talked to uses operator-furnished transportation, catering, and accommodations. The California State Lands Commission made a point of saying that being able to talk to the crew when its personnel travel to an installation in operator-furnished transportation is extremely beneficial in determining what is really going on at the installation. All regulatory bodies consulted by the committee agreed that time spent offshore in operator-furnished accommodations is essential to understanding the culture of safety on the facility. BSEE inspectors currently spend an extensive amount of time traveling to and from offshore installations. A more efficient use of available manpower would be to use the extensive amount of travel time to and from offshore

---

[1] P. Schroeder. Pacific OCS Regional Office, BSEE. Presentation to a subgroup of the committee, Camarillo, California, March 22, 2010.

[2] Inspectors do stay overnight on some facilities that are very far from land.

installations for informal discussions with operator personnel before a formal audit or visit to an offshore installation.

Furthermore, BSEE has a finite budget, and that budget should be maximized so that the inspectors' work can be fully effective. A large portion of BSEE's budget is allocated to offshore transportation costs, and the rules BSEE inspectors must adhere to became more stringent following the Macondo well accident. It would be beneficial to identify a way to minimize these costs and reallocate some of these resources to the hiring and retaining of highly capable staff. BSEE should aspire to having its inspectors and engineers be recognized as being among the more highly qualified people in the offshore industry.

Therefore, to the maximum extent practicable consistent with its increasing safety, environmental, and auditing responsibilities, BSEE, with industry input, should analyze the benefits and risks of using operator-furnished transportation and accommodations when performing inspections and audits. The agency should plan its budget recognizing that per dollar spent, the safety value of ensuring that high-quality inspectors and auditors are recruited and appropriately compensated and that critical data are collected, stored, and analyzed is greater than maintaining a completely independent transportation capability. Allowing overnight stays would increase the time BSEE staff would be able to spend interacting with the operating crew. More time on an installation would enable inspectors to better judge the degree to which a safety culture exists there. All other offshore regulators that the committee heard from believed that these measures would lead to better inspections and a higher degree of operating safety and, with proper management, would not lead to conflicts of interest. **In the analysis, consideration should also be given to increasing the fees charged for inspections and to changing the structure by which inspection (and audit) costs are passed on to industry.**

Finally, as noted in Conclusion 4, operating conditions, personnel, production requirements, and technologies are continually changing. Therefore, **BSEE should design and implement its inspection program on the basis of risk.** The use of a risk-informed framework that focuses attention where new or potential problems are likely to occur will aid in the evolution of practices and audit procedures.

Operator and BSEE Audits

Besides the inspections described above—which are, in essence, checks of compliance with specific regulations—and spot checks to determine whether individual elements of SEMS are being used effectively, a system for evaluating the effectiveness of an operator's SEMS program requires routine periodic as well as incident- and event-driven audits. For these audits to evaluate the effectiveness of a SEMS program successfully, auditors will need to understand how the organization's safety culture is reflected in its implementation of its SEMS program. Developing this understanding will require auditors to interact with operating crews and ask questions pertinent to how well crew members understand the SEMS program and how well used the program's elements are in day-to-day practice. As noted in Chapter 3, techniques used in inspections (e.g., interviews and witnessing) should also be used when conducting audits.

BSEE is responsible for ensuring that audits are conducted in a timely fashion, are thorough, and accomplish the goals set out above. **The audit scheme adopted by BSEE should have the following characteristics: operator ownership, audit team independence, training and accreditation of auditors, access to top levels of management and audit reports, a definition of required audit frequency, and a scheme for quality assurance of audits.** BSEE will also need to audit operator audit reports and have personnel capable of carrying out these tasks effectively. **Except in the case of highly deficient systems, the goal of the audit should not be to pass or to fail. Rather, an audit should be designed to help senior management by presenting them with an independent view of the state of their efforts to establish a safety management system and, by extension, a culture of safety.** The audit should identify areas for improvement and measure progress toward improvements recommended in previous audits. In its program, BSEE should take into account that safety management is a dynamic process that evolves with time and that should not be judged solely on a pass–fail system.

Operator Ownership
BSEE should ensure that operators are involved in the audit itself. Several members of the committee have participated in financial, quality, and safety audits and have observed that properly conducted inter-

nal audits by personnel familiar with the operation are much better at uncovering problems than are external audits. In discussions with the committee, a representative of the Center for Offshore Safety (COS) stated that his company operated on this principle in its safety audits and that if BSEE mandated third-party audits, his company would merely add the third-party audit as a regulatory requirement to the internal audits it was already conducting. His company would not accept the third-party audit as a substitute for one of its internal audits.

Large operators such as ExxonMobil and Shell have the ability to form independent audit teams within their organization. There are many smaller operations, however, that do not have a pool of skilled personnel from outside the operating organization being audited who are capable of performing an adequate audit of the organization. These operators should be able to use third-party auditors. **If an independent third party must be used, at least one member of the audit team should be from the operator's organization, and that person should not be directly involved in the day-to-day operation being audited.** In some cases of very small operators with only a handful of employees, it may be necessary for the chief executive officer of the company to participate as a member of the audit team.

Although operators are responsible for conducting audits, BSEE is responsible for verifying that quality audits are carried out and acted on appropriately. **BSEE should perform complete or partial audits of SEMS programs when justified by reports from inspectors, reviews of operators' audit reports, incidents, or events.** BSEE's oversight of internal and third-party audits needs to include a range of techniques, each of which focuses on a different aspect of an operation's safety system. **BSEE can use reports from its compliance inspectors and its reviews of audit reports to identify the need for specific BSEE-conducted targeted or spot audits, or complete audits, to determine whether an operator's SEMS program is improving safety.** Direct spot inspections to verify that specific requirements are being met could perhaps be accomplished by relying on checklists. **BSEE can also check to determine whether an organization's SEMS program is improving its safety culture.** Interviews, demonstrations, and observations, rather than checklists, are necessary to make such a determination. For example, the question, "Do you have an MOC (management of change process)?" would shift to, "Show me how you know if your MOC is working properly." Operators with an effective safety culture will be able to answer that question—and, in fact, describe the possible weaknesses in the MOC as implemented—even if it meets the letter of law.

Audit Team Independence
The operator's audit of its SEMS program should be conducted by a truly independent, qualified team. It is critical that the audit team be made up of members who are divorced from the organization within the company that is responsible for the day-to-day operations of the installation and for meeting the financial, operational, safety, or environmental targets set by management. **The audit team should report to the highest level practical given the size and complexity of the operator's organization.** Members of the audit team could be permanently assigned or assigned on a rotating basis for a set number of years. The latter method helps disseminate information about and respect for the audit team throughout the organization. Participation in the audit team may also be considered as an interim assignment to higher level operations assignments within the company. In no instance should the audit team include as a member someone who was recently assigned to the offshore facility being audited.

Training and Accreditation of Auditors
Audit team members should be trained to conduct audits and should be accredited by a method prescribed by BSEE.[3] General qualifications for SEMS auditors are described in more detail in Chapter 5.

BSEE must have independence (from industry) in how it trains its auditors. Nonetheless, the agency should consider certifying its auditors using a process similar to that used to certify industry auditors, and the certification should be of the same standard as outside accreditation institutions. **BSEE, in consultation with the industry and, potentially, COS, should develop an approach to certify auditors, develop audit standards, and establish the process by which audits themselves are conducted.**

---

[3] This recommendation is supported by the National Academy of Engineering–National Research Council (NAE-NRC) report on the Macondo Well–*Deepwater Horizon* blowout (NAE-NRC 2011).

Access to Top Levels of Management and Audit Reports

The key to a successful audit system lies in discussions with top management and in the steps top management takes toward continuous improvement. **BSEE should ensure that the audit team has reviewed the audit report with top levels of management and obtained their sign-off on findings and areas for improvement. A copy of the audit report and a summary of its findings and conclusions should be sent to BSEE so that the agency can spot trends and disseminate information to the industry in a timely manner.**

As an alternative, BSEE could consider allowing COS to screen all reports. Doing so would bring in an element of the peer-review–peer-assist method for assessing effectiveness that is described in Chapter 3 and would further involve the industry as a whole in taking ownership of the development of a culture of safety. Such a charge from BSEE to COS would be consistent with the following elements of the COS operating basis as presented to the committee:[4]

- Compiling and analyzing key industry metrics,
- Sponsoring functions to facilitate sharing and learning,
- Identifying and promoting opportunities for the industry to continuously improve,
- Interfacing with industry leaders to ensure leadership and system deficiencies are recognized and addressed promptly, and
- Communicating with government and external stakeholders.

Audit Frequency

Under 30 CFR 250, Subpart S, as it currently stands, the timing for audits is very prescriptive. An operator must audit every element in its SEMS program every 3 years and include at least 15 percent of its installations.

Installations in the Gulf of Mexico are very diverse. There are single-well unmanned installations, manned and unmanned installations with production equipment and no wells, manned and unmanned installations with both wells and production equipment, platforms with simultaneous drilling and production operations, floating and bottom-supported

---

[4] J. Toellner. Center for Offshore Safety. Presentation to the committee, Houston, Texas, October 19, 2011.

platforms, platforms producing 200 barrels of oil a day and platforms producing 200,000 barrels of oil a day, and all manner of drilling and workover rigs. Some installations produce high-pressure oil, which has the potential of flowing at high rates to the surface; others require pumps to lift oil to the surface. Some facilities produce natural gas with few impurities; others produce gas that contains levels of acid gases such as carbon dioxide and hydrogen sulfide. Some drilling operations are conducted in well-defined subsurface environments, while others are geologically uncertain. Some facilities have old equipment; others are new.

Similarly, operators in the Gulf of Mexico are very diverse. Some are large, international, integrated companies; others are large, domestic energy and petroleum companies; and still others are very small independent operators with only minimal staff. Some operators are responsible for a large number of fields and installations, and some operate only one or two fields and installations. Some fields have multiple platforms tied together by pipelines or bridges, or both, and some have just one platform.

Thus, it is difficult to establish a formula for audit frequency that does not become a paperwork burden and exercise for some operations while it is at the same time too lenient an audit frequency for others. Neither result is conducive to using SEMS to help establish an improved culture of safety in the industry.

Because of the diversity of operations and operators, each operator should be allowed to develop its own audit plan, subject to BSEE approval. Operator development of the audit plan would be a further step in establishing operator ownership in SEMS and its implementation and would replace the current prescriptive frequency with a more appropriate risk-based audit frequency.

Quality Assurance of Audits

In any system involving audits, BSEE is responsible for monitoring the quality of the audits and for ensuring that what is learned from the audits is implemented. Under 30 CFR 250, Subpart S, as it currently stands, BSEE accomplishes this task by requiring that operators submit their audit plans before conducting audits, submit the qualifications of audit team members and third-party audit companies, and submit the results of the audits once they have been completed. The assumption is that BSEE will review and approve all submittals and will disseminate to the industry what is learned from the audits. **In addition to requiring operators to submit an audit plan to BSEE, the agency should further require them to identify and report the follow-up actions taken as a result of the audit.** BSEE should require that the frequency and scope of the audits specified in an operator's audit plan be guided by risk rather than by a one-size-fits-all formula.

Auditing Audits
In addition to conducting spot inspections of SEMS compliance, BSEE should audit the quality of an operator's audits. This task includes performing spot inspections of documentation and audits and, where appropriate, more complete BSEE reaudits of specific facilities. BSEE should also have a plan for carrying out these activities. **In its audits, BSEE should use objective and subjective risk-based processes such as those employed by Petroleum Safety Authority (PSA) Norway, and these audits should be carried out by BSEE employees who are themselves accredited.**

Personnel for Auditing
BSEE should hire or train a sufficient number of auditors, including qualified audit team leaders and an adequate number of staff for analyzing audit reports effectively and auditing the accreditation system that the agency puts in place.

Key Performance Indicators

Specific KPIs associated with SEMS effectiveness are difficult to define and need further study and evaluation by both the industry and BSEE. Common safety and environmental metrics such as the number of injuries per year or the volume of spills per year provide only a part of the effectiveness picture. Other metrics need to be identified as lagging or leading indicators in relation to process safety. Once identified, such metrics can be used to monitor and direct the improvement of SEMS.

BSEE can collect and evaluate data from operations within and across installations to identify specific problems and trends in operations at a particular facility and across the industry. This information is needed to evaluate the SEMS audit approach and to identify opportunities for improvement. While the benefits from such a data exchange are obvious and important, implementation is far from trivial. An open data-collection and data-sharing protocol requires agreements across all parties to ensure that confidentiality and legal concerns are satisfied. **BSEE should distribute information in a timely manner to the industry on trends and methods for improving the SEMS process and overall safety, as well as lessons learned, by means of publications, workshops, seminars, and other methods.**

Offshore safety organizations abroad that have programs similar to SEMS, such as PSA Norway and the United Kingdom (UK) Health and Safety Executive (HSE), have access to a considerable amount of data. Because many of the safety and environmental issues associated with offshore oil and gas operations are common worldwide, a data set compiled from all of these organizations would be invaluable. **BSEE should create a task force with the industry, PSA Norway, the UK HSE, and other similar regulatory bodies worldwide to identify KPIs.** Creation of such a task force will help BSEE ensure that it is collecting the proper SEMS-relevant data and analyzing it appropriately to direct the agency's effort to measure the effectiveness of SEMS.

Whistleblower Program

PSA Norway, the Occupational Safety and Health Administration, and other organizations have found that programs that allow personnel to anonymously report possible violations directly to the regulator are helpful in identifying possible issues that may not be found by other means. The SEMS II notice of proposed rulemaking (BOEMRE 2011a) describes an approach that provides for anonymous reports of potential violations. **BSEE should have a program for anonymous reporting and a process to follow up such reports and should use the information gained from them appropriately to modify BSEE inspections and audits.** This program should also allow for the anonymous reporting of inappropriate behavior of BSEE personnel and potential improvements in BSEE policies and procedures, as well as potential violations by operators. Care should be taken in devising the program to make sure that it does not become a tool for disgruntled employees seeking to punish perceived wrongs. This recommendation supports Summary Recommendations 5.4 and 6.14 in the NAE-NRC report on the Macondo well–*Deepwater Horizon* blowout:

> Industry, BSEE, and other regulators should improve corporate and industrywide systems for reporting safety-related incidents. Reporting should be facilitated by enabling anonymous or "safety privileged" inputs. Corporations should investigate all such reports and disseminate their lessons-learned findings in a timely manner to all their operating and decision-making personnel and to the industry as a whole. A comprehensive lessons-learned repository should be maintained for industrywide use. This information can be used for training in accident prevention and continually improving standards. (NAE-NRC 2011, pp. 107 and 123)

RESOURCES REQUIRED

BSEE should analyze its budget to ensure that it has sufficient financial resources to implement these recommendations. Savings from any increased use of operator transportation and more efficient time offshore derived from operator-furnished accommodations could potentially be reprogrammed for the agency's enhanced inspection and SEMS audit programs and other related activities required by these recommendations.

References

Abbreviations

ABS	American Bureau of Shipping
API	American Petroleum Institute
BOEMRE	Bureau of Ocean Energy Management, Regulation, and Enforcement
BSI	British Standards Institution
CAIB	*Columbia* Accident Investigation Board
CSB	U.S. Chemical Safety Board
DNV	Det Norske Veritas
HSE	Health and Safety Executive, United Kingdom
IADC	International Association of Drilling Contractors
ISO	International Organization for Standardization
MMS-USCG	Minerals Management Service–U.S. Coast Guard
MOC	Management of Change
NAE-NRC	National Academy of Engineering–National Research Council
NRC	National Research Council
OECD	Organisation for Economic Co-operation and Development
PSA	Petroleum Safety Authority
SPE	Society of Petroleum Engineers
USCG	U.S. Coast Guard
U.S. NRC	U.S. Nuclear Regulatory Commission

ABS. 2012. *Guidance Notes on Safety Culture and Leading Indicators of Safety.* Houston, Tex. http://www.eagle.org/eagleExternalPortalWEB/ShowProperty/BEA%20Repository/Rules&Guides/Current/188_Safety/Guide. Accessed May 31, 2012.

ABSG Consulting Inc. 2006. *Review of Process Safety Management Systems at BP North American Refineries for the BP U.S. Refineries Independent Safety Review Panel.* ABSG Consulting Inc., Houston, Tex.

API. 1990. *Management of Process Hazards.* API RP 750 (with errata). API, Washington, D.C.

API. 1993. *Recommended Practice for Development of a Safety and Environmental Management Program for Offshore Operations and Facilities,* 1st ed. API RP 75. API, Washington, D.C.

API. 2003. *Management of Hazards Associated with Location of Process Plant Buildings,* 2nd ed. API RP 752. API, Washington, D.C.

API. 2004. *Recommended Practice for Development of a Safety and Environmental Management Program for Offshore Operations and Facilities,* 3rd ed. API RP 75. API, Washington, D.C.

API. 2010a. *Fatigue Risk Management Systems for Personnel in the Refining and Petrochemical Industries.* API RP 755. API, Washington, D.C.

API. 2010b. *Process Safety Performance Indicators for the Refining and Petrochemical Industries.* API RP 754. API, Washington, D.C.

Appleton, B. 1995. Lessons in Safety Management. Video. Mobil North Sea Technology Forum, 1995.

Bea, R. 2002. Human and Organizational Factors in Reliability Assessment and Management of Offshore Structures. *Risk Analysis,* Vol. 22, No. 2, pp. 19–34.

BOEMRE. 2010. Oil and Gas and Sulphur Operations in the Outer Continental Shelf—Safety and Environmental Management Systems. *Federal Register,* Vol. 75, No. 199, Fri., Oct. 15, 2010, pp. 63610–63654.

BOEMRE. 2011a. Oil and Gas and Sulphur Operations in the Outer Continental Shelf—Revisions to Safety and Environmental Management Systems. *Federal Register,* Vol. 76, No. 178, Wed., Sept. 14, 2011, pp. 56683–56694.

BOEMRE. 2011b. *Report Regarding the Causes of the April 20, 2010 Macondo Well Blowout.* http://www.boemre.gov/pdfs/maps/dwhfinal.pdf.

Booth, R. 1993. Safety Culture: Concept, Measurement and Training Implications. *Proceedings of the British Health and Safety Society Spring Conference: Safety Culture and the Management of Risk,* Apr. 19–20, 1993.

Borthwick, D. 2010. *Report of the Montara Commission of Inquiry.* Montara Commission of Inquiry, Commonwealth of Australia. http://www.ret.gov.au/Department/Documents/MIR/Montara-Report.pdf. Accessed Mar. 3, 2011.

BP U.S. Refineries Independent Safety Review Panel. 2007. *The Report of the BP U.S. Refineries Independent Safety Review Panel* (also referred to as the "Baker Panel Report"). http://www.bp.com/liveassets/bp_internet/globalbp/globalbp_uk_english/SP/STAGING/local_assets/assets/pdfs/Baker_panel_report.pdf.

BSI. 2012. *Guidelines for Auditing Management Systems (ISO 19011:2011).* BS EN ISO 19011:2011. British Standards Institution, London.

CAIB. 2003. Organizational Causes: Evaluating Best Safety Practices. In *The CAIB Report: Volume 1,* pp. 182–184. http://caib.nasa.gov/news/report/pdf/vol1/full/caib_report_volume1.pdf.

CSB. 2007. *Investigation Report: Refinery Explosion and Fire (15 Killed, 180 Injured).* Report No. 205-04-I-TX. BP, Texas City, Tex., Mar. 23, 2005. http://www.csb.gov/assets/document/CSBFinalReportBP.pdf. Accessed Apr. 4, 2012.

Cullen, W. 1990. *The Public Inquiry into the* Piper Alpha *Disaster*. HM Stationery Office, London.

DNV. 2011. *Major Hazard Incidents Arctic Offshore Drilling Review*. Prepared for the National Energy Board. Report No. NEB 2010-04/DNV, Reg. No. ANECA 851. DNV, Houston, Tex.

Hopkins, A. 2004. *Safety, Culture, and Risk: The Organisational Causes of Disasters*. CCH Australia Limited, Sydney, Australia.

Hopkins, A. 2006. *Working Paper 44: Studying Organisational Cultures and Their Effects on Safety*. National Research Centre for OAS Regulation, Australian National University, Canberra, Australian Capital Territory. http://regnet.anu.edu.au/sites/default/files/u86/WorkingPaper_44.pdf.

Hopkins, A. 2009. *Failure to Learn: The BP Texas City Refinery Disaster*. CCH Australia Limited, Sydney, Australia, 2009.

HSE. 2011. Offshore Research. http://www.hse.gov.uk/offshore/offshoreresearch.htm. Accessed Apr. 4, 2011.

IADC. 2011. Ralls: Industry Safety Culture Is Not Complacent. *Drilling Contractor*, Feb. 8, 2011. http://www.drillingcontractor.org/ralls-industry-safety-culture-is-not-complacent-8482.

ISO. 2002. *ISO 19011:2002. Guidelines for Quality and/or Environmental Management Systems Auditing*. Geneva, Switzerland. http://www.iso.org/iso/catalogue_detail?csnumber=31169.

ISO. 2005. *ISO 9000:2005. Quality Management Systems—Fundamentals and Vocabulary*. Geneva, Switzerland. http://www.iso.org/iso/iso_catalogue/catalogue_tc/catalogue_detail.htm?csnumber=42180.

ISO. 2008. *ISO 9001:2008. Quality Management Systems—Requirements*, 4th ed. Geneva, Switzerland. http://www.iso.org/iso/iso_catalogue/catalogue_tc/catalogue_detail.htm?csnumber=46486.

LaPorte, T. R., and P. M. Consolini. 1991. Working in Practice but Not in Theory: Theoretical Challenges of "High-Reliability" Organizations. *Journal of Public Administration Research and Theory*, Vol. 1, No. 1, pp. 19–47. http://www.cs.st-andrews.ac.uk/~ifs/Teaching/Socio-tech-systems(LSCITS)/Reading/HROs.pdf.

Leveson, N. G. 2011. *Engineering a Safer World: System Thinking Applied to Safety*. Massachusetts Institute of Technology Press, Cambridge.

Meshkati, N. 1999. Cultural Context of Nuclear Safety Culture: A Conceptual Model and Field Study. In *Nuclear Safety: A Human Factors Perspective* (J. Misumi, B. Wipert, and R. Miller, eds.), McGraw-Hill, New York, pp. 113–130.

MMS-USCG. 2004a. Memorandum of Agreement Between the Minerals Management Service, U.S. Department of the Interior and the U.S. Coast Guard, U.S. Department of Homeland Security. Subject: Agency Responsibilities. MMS/USCG MOA: OCS-

01. Sept. 30, 2004. http://www.uscg.mil/hq/cg5/cg522/cg5222/docs/mou/AGENCY_RESPONSIBILITIES.pdf.

MMS-USCG. 2004b. *Memorandum of Understanding Between the Minerals Management Service, U.S. Department of the Interior and the U.S. Coast Guard, U.S. Department of Homeland Security.* Sept. 30, 2004. http://www.boemre.gov/regcompliance/MOU/cgmoufnl.htm.

MMS-USCG. 2006a. *Memorandum of Agreement Between the Minerals Management Service, U.S. Department of the Interior and the U.S. Coast Guard, U.S. Department of Homeland Security. Subject: Civil Penalties. MOA: OCS-02.* September 12, 2006. http://www.uscg.mil/hq/cg5/cg522/cg5222/docs/mou/CIVIL_PENALTIES.pdf.

MMS-USCG. 2006b. *Memorandum of Agreement Between the Bureau of Safety and Environmental Enforcement, U.S. Department of the Interior, and the U.S. Coast Guard, U.S. Department of Homeland Security. Subject: Oil Discharge Planning, Preparedness, and Response. MOA: OCS-03.* Apr. 3, 2012. http://www.uscg.mil/hq/cg5/cg522/cg5222/docs/mou/MOA%20OCS%2003_FINAL_Signed_3APR12.pdf.

MMS-USCG. 2008. *Memorandum of Agreement Between the Minerals Management Service, U.S. Department of the Interior and the U.S. Coast Guard, U.S. Department of Homeland Security. Subject: Floating Offshore Facilities. MOA: OCS-04.* Feb. 28, 2008. http://www.uscg.mil/hq/cg5/cg522/cg5222/docs/mou/FLOATING_OFFSHORE_FACILITIES.pdf.

NAE-NRC. 2011. *Macondo Well–Deepwater Horizon Blowout: Lessons for Improving Offshore Drilling Safety.* National Academies Press, Washington, D.C.

National Commission on the BP *Deepwater Horizon* Oil Spill and Offshore Drilling. 2011. *Deep Water: The Gulf Oil Disaster and the Future of Offshore Drilling.* Report to the President. http://www.oilspillcommission.gov/sites/default/files/documents/DEEPWATER_ReporttothePresident_FINAL.pdf.

Norwegian Petroleum Directorate. 2011. *Facts 2011—The Norwegian Petroleum Sector.* Publication No. Y-0103 E. Ministry of Petroleum and Energy, Stavanger, Norway. http://www.npd.no/en/Publications/Facts/Facts-2011.

NRC. 1990. *Alternatives for Inspecting Outer Continental Shelf Operations.* National Academies Press, Washington, D.C.

OECD. 1999. *The Role of the Nuclear Regulator in Promoting and Evaluating Safety Culture.* Nuclear Energy Agency, OECD, Paris. http://www.oecd-nea.org/nsd/reports/nea1547-Murley.pdf.

Peters, T., and R. Waterman. 1982. *In Search of Excellence: Lessons Learned from America's Best-Run Companies.* Harper & Row, New York.

PSA Norway. 2002. *White Paper No. 7 (2001–2002): On Health, Safety and the Environment in Petroleum Operations.* PSA Norway, Stavanger.

PSA Norway. 2010. *From Prescription to Performance in Petroleum Supervision.* PSA Norway, Stavanger. http://www.ptil.no/news/from-prescription-to-performance-in-petroleum-supervision-article6696-79.html.

PSA Norway. 2011a. *Assessments and Recommendations after* Deepwater Horizon. PSA Norway, Stavanger. http://www.ptil.no/news/assessments-and-recommendations-after-deepwater-horizon-article7890-79.html?lang=en_US.

PSA Norway. 2011b. *The* Deepwater Horizon *Accident—Assessments and Recommendations for the Norwegian Petroleum Industry.* PSA Norway, Stavanger. http://www.ptil.no/getfile.php/PDF/DwH_PSA_summary.pdf.

PSA Norway. 2011c. *Safety Status & Signals 2010–2011.* PSA Norway, Stavanger.

PSA Norway. 2011d. *Supervision: A Finger on the Pulse.* PSA Norway, Stavanger. http://www.ptil.no/supervision/supervision-a-finger-on-the-pulse-article7624-88.html.

Rasmussen, J. 1997. Risk Management in a Dynamic Society: A Modeling Problem. *Safety Science,* Vol. 27, No. 2–3, pp. 183–213.

Rasmussen, J., and I. Svedung. 2000. *Proactive Risk Management in a Dynamic Society.* Swedish Rescue Services Agency, Karlstad. https://www.msb.se/RibData/Filer/pdf/16252.pdf. Accessed Apr. 4, 2012.

Reason, J. 1983. Achieving a Safe Culture: Theory and Practice. *Work and Stress,* Vol. 12, No. 3, pp. 293–306.

Reason, J. 1997. *Managing the Risks of Organizational Accidents.* Ashgate Publishing Company, Aldershot, UK.

Scarlett, L., I. Linkov, and C. Kousky. 2011. *Risk Management Practices: Cross-Agency Comparisons with Mineral Management Service.* Discussion Paper # RPP DP 10-67. Resources for the Future, Washington, D.C.

Schein, E. H. 1992. *Organizational Culture and Leadership,* 2nd ed. Jossey-Bass, San Francisco, Calif.

Schein, E. H. 2004. *Organizational Culture and Leadership,* 3rd ed. Jossey-Bass, San Francisco, Calif.

USCG. 2011. *Report of Investigation into the Circumstances Surrounding the Explosion, Fire, Sinking and Loss of Eleven Crew Members Aboard the Mobile Offshore Drilling Unit* Deepwater Horizon *in the Gulf of Mexico April 20–22, 2010, Vol. I.* MISLE Activity Number: 3721503. https://www.hsdl.org/?view&did=6700.

U.S. NRC. 2011. *Safety Culture Policy Statement.* NUREG/BR-0500. http://pbadupws.nrc.gov/docs/ML1116/ML11165A021.pdf. Accessed Apr. 4, 2012.

Weick, K. E. 1987. Organizational Culture as a Source of High Reliability. *California Management Review,* Vol. 24, pp. 112–127.

Weick, K. E., and K. M. Sutcliffe. 2001. *Managing the Unexpected: Assuring High Performance in an Age of Complexity.* Jossey-Bass, San Francisco, Calif.

Weick, K. E., and K. M. Sutcliffe. 2007. *Managing the Unexpected: Resilient Performance in an Age of Uncertainty,* 2nd ed. Jossey-Bass, San Francisco, Calif.

Study Committee Biographical Information

Kenneth E. Arnold (Member, National Academy of Engineering), *Chair*, is a senior technical advisor for WorleyParsons with more than 45 years of experience in projects, facilities, and construction related to upstream oil and gas development. He spent 16 years at Shell in engineering and engineering and research management before forming Paragon Engineering Services, a project management and offshore engineering company, in 1980; it had a staff of 600 when it was sold to AMEC in 2005. Mr. Arnold is the author, coauthor, or editor of several textbooks and numerous technical articles on the design and project management of production facilities. He taught production facility design at the University of Houston and has been active in the Society of Petroleum Engineers (SPE) and other technical societies. He was named Houston's 2003 Engineer of the Year by the Texas Society of Professional Engineers and is the recipient of the SPE Public Service Award and the DeGolyer Distinguished Service Medal. He was elected to the National Academy of Engineering in 2005, primarily for the work he has done in promoting offshore safety. Mr. Arnold has served on two Marine Board committees, including the 1990 Committee on Alternatives for Offshore Inspection, and was a member of the Marine Board for 6 years.

J. Ford Brett is managing director of PetroSkills and chief executive officer of Oil and Gas Consultants International (OGCI), the world's largest petroleum training organization. Mr. Brett has consulted in more than 25 countries worldwide in the area of petroleum project and process management. Before joining OGCI, he was with Amoco Production Company, where he worked on drilling projects in the Bering Sea, the North Slope of Alaska, the Gulf of Mexico, offshore Trinidad, and

Wyoming. In 2000, the American Society for Competitiveness awarded him the Crosby Medallion for Global Competitiveness for work in "global competitiveness through quality in knowledge management, best practices transfer, and operations improvement." He currently serves on the board of the Society of Petroleum Engineers as technical director for drilling and completion. For his work on improved drilling techniques, he was also honored in 1996 with a nomination for the National Medal of Technology, the U.S. government's highest technology award. Mr. Brett has been granted more than 25 U.S. and international patents and has authored or coauthored more than 25 technical publications. He holds a BS degree in mechanical engineering and physics from Duke University, an MSE from Stanford University, and an MBA from Oklahoma State University.

Paul S. Fischbeck is professor in the Department of Engineering and Public Policy and the Department of Social and Decision Sciences at Carnegie Mellon University. He is also director of the Carnegie Mellon Center for the Study and Improvement of Regulation, where he coordinates a diverse research group exploring all aspects of regulation, from historical case studies to transmission-line siting to emissions-trading programs. Widely published, Dr. Fischbeck has served on a number of national research committees and review panels, including the National Research Council (NRC)–Transportation Research Board (TRB) Committee on School Transportation Safety; the National Science Foundation's Decision, Risk, and Management Sciences Proposal Review Committee and Small Business Innovative Research Proposal Review Committee; the NRC-TRB Committee on Evaluating Double Hull Tanker Design Alternatives; and the NRC-TRB Committee on Risk Assessment and Management of Marine Systems. His research involves normative and descriptive risk analysis, including development of a risk index for prioritizing inspections of offshore oil production platforms; an engineering and economic policy analysis of air pollution from international shipping; a large-scale probabilistic risk assessment of the space shuttle's tile protection system; and a series of expert elicitations involving a variety of topics, including environmental policy selection, travel risks, and food safety. Dr. Fischbeck is cofounder of the Western Pennsylvania Brownfields Center at Carnegie Mellon, an interdisciplinary research group investigating ways to improve industrial site reuse. He is involved with a number of professional research organizations, including the American Society for Engineering Education, the Institute for Operations Research and Management Sciences, the Military Operations Research Society, and the Society of Risk Analysis. He has chaired a National Science Foundation panel on urban interactions and currently serves on the Environmental Protection Agency's Science Advisory Board. Dr. Fischbeck holds a BS in architecture from the University of Virginia, an MS in operations research and management science from the Naval Postgraduate School, and a PhD in industrial engineering and engineering management from Stanford University.

Stuart Jones is a project manager with Lloyd's Register EMEA, Aberdeen, Scotland, United Kingdom, where he is responsible for several integrity management contracts for clients operating oil and gas installations in the North Sea. He started offshore work in the oil and gas industry in 1983, when he joined Conoco in Aberdeen as its maintenance coordinator for corrosion, responsible for fabric maintenance, inspection, and corrosion monitoring on the *Murchison* and *Hutton* tension-leg platforms. He was corrosion and inspection engineer for the British Gas Rough Field operation between 1990 and 1995, when he left to follow a career more aligned with risk-based inspection. He has performed risk-based inspection studies on oil and gas installations and their associated pipelines both on- and offshore. In 2000 he joined Lloyd's Register and since then has performed a number of roles, including senior corrosion engineer, team leader, project manager, and now project controls manager. In 2009, at the initiation of this committee study, he was on a long-term international assignment with Lloyd's Register Capstone, initially as head of its Upstream Operations Team and later as head of its project controls group. He returned to work in the United Kingdom in October 2010. Mr. Jones has published a number of papers and made numerous presentations on corrosion and risk-based inspection, and from 2008 to 2010 he served on the Society of Petroleum Engineers Gulf Coast Section, Projects, Facilities, and Construction Study Group. Mr. Jones earned a second-class honors degree in metallurgy from the University College of Swansea, Wales, United Kingdom, in 1974. He is a professional member of the Institute of Corrosion and of the Institute of Materials, Minerals, and Mining and is a chartered engineer.

Thomas Kitsos served as executive director of the U.S. Commission on Ocean Policy (USCOP) from 2001 to 2004. In 2005, Dr. Kitsos retired from the National Oceanic and Atmospheric Administration, U.S. Department of Commerce, as associate deputy assistant administrator for ocean services. He is currently a private consultant on national ocean policy, advising the Joint Ocean Commission Initiative, the follow-up, foundation-supported organization composed of the members of USCOP and the privately funded Pew Ocean Commission and dedicated to promoting ocean policy reform proposals recommended by the two commissions. His earlier experience included 6 years at the U.S. Department of the Interior (DOI), where his primary responsibilities were in the area of energy development on the Outer Continental Shelf. He also served as special assistant to the assistant secretary, Land and Minerals Management, and as DOI's acting director of the Minerals Management Service, among other positions. Before his tenure at DOI, Dr. Kitsos spent 20 years on Capitol Hill on the staff of the U.S. House of Representatives Committee on Merchant Marine and Fisheries. His final position with the committee was as chief counsel, advising the chairman on national ocean and coastal issues, offshore energy development, and environmental and other marine management legislation, including amendments to the Outer Continental Shelf Lands Act and the Coastal Zone Management Act. He holds BS degrees in education and social science from the Eastern Illinois University and an MA and PhD in political science from the University of Illinois.

Frank J. Puskar is managing director of Energo Engineering in Houston, Texas. Energo specializes in advanced structural engineering and structural integrity management (SIM) of existing offshore structures. Mr. Puskar has more than 28 years of experience in the offshore industry and is a recognized leader in SIM technology. He has been involved in the planning of above-water and below-water inspections and structural assessments for more than 250 fixed and floating platforms located worldwide. He has served on committees or task groups of the American Petroleum Institute (API), International Standards Organization, and American Society of Civil Engineers and on the Offshore Operators Committee and was Chairman of the API Task Group that developed API Bulletin 2HINS, *Guidance for Post-hurricane Structural Inspection* *of Offshore Structures,* published in May 2009. In 2007, he was awarded the Minerals Management Service Corporate Leadership Award for his industry efforts, including improving codes and standards related to the damage and destruction of platforms in the Gulf of Mexico from Hurricanes Ivan, Katrina, and Rita. He holds an MEng in ocean engineering from the University of California, Berkeley, and a BS in civil engineering from the State University of New York at Buffalo. He is a registered professional engineer in California, Louisiana, and Texas.

Darin W. Qualkenbush is a Health, Environment and Safety Regulatory Specialist with Chevron North America Exploration and Production Company. He served in the U.S. Coast Guard (USCG) for 24 years; his final assignment was in the National Technical Advisor office of the Outer Continental Shelf National Center of Expertise in Morgan City, Louisiana. This office is responsible for revitalizing the technical competency and expertise within the USCG marine safety program to keep pace with the growth and complexity of the offshore maritime industry. Additional duties included directing the generation of regulations, policy, and doctrine for marine safety and offshore operations as well as being a repository for USCG expertise and best practices for the offshore oil and gas industry. Lt. Qualkenbush's previous assignment was as chief, Outer Continental Shelf inspections, at the Marine Safety Unit, Morgan City, where he was responsible for all regulatory and compliance issues for exploration, exploitation, and production of oil and natural gas within USCG's approximately 69,000-square-mile offshore area of responsibility. He is a subject matter expert on lifesaving and firefighting equipment and deployment and on USCG regulatory compliance and International Maritime Organization Convention compliance on offshore oil and gas production platforms, offshore drilling units, and oil field support vessels of all types.

Raja V. Ramani (Member, National Academy of Engineering), is emeritus George H., Jr., and Anne B. Deike Chair of Mining Engineering and professor emeritus of mining and geo-environmental engineering at Pennsylvania State University, where he has been on the faculty since 1970. He is a certified first-class mine manager under the Indian Mines Act of 1952 and has been a registered professional engineer in the Commonwealth of Pennsylvania since 1971. Dr. Ramani's research activities include mine health, safety, productivity, environment, and management; flow mechanisms of air, gas, and dust in mining environs; and innovative mining methods. He has been a consultant to the United Nations, World Bank, National Safety Council, mining companies, and governmental agencies. He has published extensively on health, safety and environmental planning, and management issues and has received numerous awards from academia and technical and professional societies. Dr. Ramani was the 1995 president of the Society for Mining, Metallurgy, and Exploration (SME). He served on the U.S. Department of Health and Human Services Mine Health Research Advisory Committee from 1991 to 1998, was the chair of the National Research Council (NRC) Committee on Post Disaster Survival and Rescue from 1979 to 1981, and was a member of the Health Research Panel of the NRC Committee on the Research Programs of the U.S. Bureau of Mines in 1994. He was a member of the U.S. Department of the Interior's Advisory Board to the Director of the U.S. Bureau of Mines in 1995 and a member of the Secretary of Labor's Advisory Committee on the Elimination of Coal Worker's Pneumoconiosis from 1995 to 1996. More recently, Dr. Ramani was a member of several NRC committees, including the Panel on Technologies for the Mining Industries (2000 to 2001), the Committee on Coal Waste Impoundment Failures and Breakthroughs (2001 to 2002), the Committee to Inform Coal Policy (2005 to 2007), and the Committee to Develop the Framework for the Evaluation of NIOSH [National Institute of Occupational Safety and Health] Research Programs (2005 to 2009), and was chair of the National Academy of Sciences Committee to Evaluate the NIOSH Mining Health and Safety Research Program (2005 to 2007). In 2002, he chaired the Pennsylvania Governor's Commission on Abandoned Mine Voids and Mine Safety that was set up immediately after the Quecreek Mine inundation incident and rescue. Dr. Ramani is a distinguished member of SME (class of 1988) and an honorary member of the American Institute of Mining, Metallurgical, and Petroleum Engineers (class of 2010). Dr. Ramani holds MS and PhD degrees in mining engineering from Pennsylvania State University.

Vikki Sanders is a consultant for JMJ Associates in Austin, Texas. She assists client organizations and project teams in creating and sustaining world-class performance through JMJ's Incident, Injury-Free, and High Performance Projects practices. She works with a variety of clients in the oil and gas industry throughout the United States and Canada. After receiving her master's degree, Ms. Sanders began working in organizational development at the Aston Centre for Effective Organisations, Birmingham, United Kingdom (UK), focusing on leadership, teamwork, and employee satisfaction. She then worked in safety management and human factors at the Health and Safety Laboratory, an agency of the UK Health and Safety Executive (HSE), where she provided technical assistance to HSE inspectors, focusing on assessment of workforce tasks in multiple industries in the United Kingdom. In 2007, Ms. Sanders moved to Houston, Texas, where she worked for Atkins Global on a variety of oil and gas projects as a human factors consultant, providing human factors assessments of control rooms and other equipment for offshore platforms. She also provided input to the safety management system integration toolkit for the marine industry. Ms. Sanders graduated in psychology with honors in 1995 from the University of Humberside, United Kingdom, and earned a master's degree in organizational psychology from the University of Nottingham, United Kingdom, in 2002.